Current and Emerging Issues in the Audiovisual Industry

Diverse and Global Perspectives on Value Creation Set

coordinated by
Nabyla Daidj

Volume 1

Current and Emerging Issues in the Audiovisual Industry

Mercedes Medina
Mónica Herrero
Alicia Urgellés

iSTE

WILEY

First published 2017 in Great Britain and the United States by ISTE Ltd and John Wiley & Sons, Inc.

ISTE Ltd
27-37 St George's Road
London SW19 4EU
UK

www.iste.co.uk

John Wiley & Sons, Inc.
111 River Street
Hoboken, NJ 07030
USA

www.wiley.com

Library of Congress Control Number: 2016958362

British Library Cataloguing-in-Publication Data
A CIP record for this book is available from the British Library
ISBN 978-1-84821-977-9

MIX
Paper from responsible sources
FSC® C013604

Contents

Foreword

The changes in the audiovisual industry are enormous, facilitated by broadband digital distribution and paid services for both linear content streams and disaggregated content. This is transforming the environment in which viewers choose to access audiovisual content, the markets for that content and the fortunes of enterprises operating in the industry.

The rapidly changing environment surrounding the audiovisual industry is characterized by significant apprehension. Broadcasters are concerned that linear television streams may be declining and asynchronous viewing will become the norm. Consumers are anxious because the technology needed is becoming more complex and are worried that they are being priced out of the market for that technology and the content it provides. Pay-television operators and digital download firms are worried because policies supporting free and public service broadcasting make investment riskier and limit their abilities to pursue innovation. Social observers are troubled by the idea that free television might disappear and only pay services remain, thus making it unavailable to significant parts of the public.

It is tempting to conceive the new opportunities and new players in the audiovisual industry as revolutionizing and destroying existing producers, channels and services, but the reality is more that of adaptation, evolution and transformation as legacy producers and broadcasters adjust to new market positions and new uses for their products. This is not the task that many established players would like to undertake, because it creates an increasingly competitive struggle for audiences and financial resources. Many wish the situation would just end and are resisting changes necessary to make their firms more competitive. Others are reducing the quality of content as resources diminish, harming themselves and the overall quality of content in society.

Some firms are struggling valiantly to make sense of the new environment and find sustainable and profitable pathways. They are trying to comprehend the new forces at work, seeking to find what they can do and making efforts to understand what changes their enterprises will be required to make. Identifying strategies that producers, broadcasters, cablecasters and other pay-television services can employ in this environment is crucial. This book helps make sense of the transformation underway and the issues that different participants in the audiovisual industry are encountering.

The authors of this book use the Spanish audiovisual industry as a case to examine and explain the nature of the challenges. They show how investing carefully, innovating processes, products and distribution, developing new relationships with audiences, creating new business opportunities and business models, and improving quality content are important elements of transformation.

Underlying all the chapters is a focus on audiences and consumers and how the additional audiovisual services and choices are altering what, when and how they watch audiovisual content as well as their willingness to pay for it. The significant shift in consumer behavior indicates that considerable portions of the public were unhappy with previous content and services, and had only few options but to consume what was offered.

In evaluating the changes, the authors describe markets and processes that many traditional broadcasters and those producing for those broadcasters do not understand well. They reveal the new economic influences and relationships and how market power is shifting in the audiovisual content from suppliers to consumers. They explore how the confusion over functions and roles of traditional television channels in the new environment is confounding their response and why it must change.

The authors show that the audiovisual industry has become much more complex and that companies must become much more aware and decisive about their place in it. They also show that existing companies need clear strategies and new objectives for coping with the increased competition and new services that attract consumer attention and payments. The authors provide some advice on how to think about the changes and develop new strategies that will be necessary.

This is a useful book that is focused on change, innovation and helping to develop thinking about how to improve the competitive positions of companies in the new environment and how to better link them to audiences and paying consumers who will be needed to improve their revenues and sustainability in the future. Solving these issues is critical if society is to have the type of audiovisual industry necessary to provide quality entertainment and informational services that serve social, cultural and political needs.

Robert G. PICARD
Reuters Institute
University of Oxford
December 2016

Introduction

The television industry is facing important structural changes worldwide, which has been in constant expansion and change since the beginning. It is a relatively new industry, naturally open to innovation and highly influenced by technology and the consequent improvements in the quality of delivery or consumption.

A problem arises before the digital challenge and the threat of new online audiovisual services. It needs to be solved by the industry, its regulators and by academic research: what will TV be and what will be its position in the value chain? Following Evans [EVA 11, p. 1], "television is now bigger than TV", but television has not died at all.

According to research carried out under the title "New habits of audiovisual consumption in Europe: impact of digitization in the industry and the media diet of citizens", funded by the Spanish Ministry of Economy and Competitiveness, in 2012, 83% of respondents preferred to see programs on television over other devices; 65.5% had no interest in watching online content, 51% had difficulty finding them and 35% had poor connection. These data show that there is still scope for linear television, with scheduled content on fixed days [MED 15].

Observation of the market in recent years has shown that television is still the favorite medium for audiovisual content. However, in the new digital era, television will have to demonstrate some of the functions for which it is irreplaceable: specifically, as promoter of live events; as discoverer,

Chapter written by Mercedes MEDINA.

developer and distributor of content and of course, producer. New media benefit from the experience of traditional television, which has been the expert for more than 60 years in gathering viewers, entertaining and exciting all audiences, both young and old [MUR 11].

Arrese [ARR 06] notes that information and entertainment products are attention goods, which require experiencing them to assess their value, and generate economies of scale as well. Therefore, its value should not be measured only by the costs of production and reproduction, but by the level of satisfaction and utility they give to the user. On the same matter, Doyle [DOY 10] notes that the success of content is measured by its ability to "hook" the audience. In order to "hook" the viewers in the new competitive environment, broadcasters should develop systems for measuring satisfaction and usefulness of the content and tools to gather better information about viewers and find an adequate supply of different profiles.

Traditional audiovisual media companies have managed to create competitive strategies in the national market and internationally. Through a process of vertical and horizontal growth, they have managed to be solid companies, as shown by their economic data. With precise programming strategies, they have achieved high audience shares. Thanks to the multiplatform strategy, they have spread their content to all devices available to users. The role that audiovisual companies will have to continue to play in the coverage of events and in the interpretation and understanding of them for the general public, as well as their role in social cohesion, is unquestionable.

The new breaks into the old but not as a complete killer. The old remains. Therefore, the order of the chapters in this book combines the new with the old as it happens in the market. In this book, we consider current and emergent issues, new entrants play the role of directors of trends and help us observe how markets and legacy media transform.

Although some examples of the Spanish market are provided, reflections and trends are valid for other countries. In today's globalized world, the echo of phenomena multiplies and has similar effects on all economies founded on free markets. Some chapters are more theoretical and others provide information on market analysis. The ideas collected here are the result of years of market observation, conversations with media managers and trend analysis. The analysis in the discipline of media economics is not understood without the use of data that reflect the behavior of markets and companies.

In Chapter 1, we reflect on the new paradigm of media that technology has instigated. The media continue to seek havens of freedom to perform the function for which they were created. In some countries, these havens exist and function, but in others, the consequences of freedom have not always been freedom for all or freedom for better products. Market concentration and loss of content quality are some of the current threats. In any case, the current challenges for media companies lie in the development of competitive content together with the successful compliance of the function to inform and entertain. Innovation in discovering alternative revenue streams has become a key issue for media managers.

Chapter 2 describes the entry of new audiovisual online services on the market that have challenged the legacy media companies and led to some paradigm shifts in traditional business models. The profile of new companies, namely Netflix, and their way of reaching audiences are a great revolution for pay-TV operators.

Chapter 3 questions the notion that new media are to eliminate their predecessors and discusses the value of the contributions of private television companies to the market. Although new competitors arrive, the bargaining power and weight in the market of audiovisual companies remain important for the media industry in most European countries.

In Chapter 4, we gaze at the increasingly difficult and less sustainable role of public television in Europe. We analyze the Spanish model which highlights the economic and financial difficulties of maintaining a corporation like RTVE and the various solutions that governments have launched.

Again, Chapter 5 questions the role of new operators after an analysis of consumer trends in recent years. In this case, the focus turns to the audience of audiovisual content. Despite the new offerings and consumption possibilities, these remain higher for traditional television systems.

In a book on the current issues and trends of the media market, academic research demands reflection on the quality of content. The content of the media is the driving force and rationale of this industry, if its content is not relevant, engaging and compelling for audiences, the whole structure of this market will collapse and will require a larger industrial structure than there has been in recent years. A view of the quality of content from the business perspective seems necessary as the foundation of media management in the coming years. In this sense, quality is an implicit commitment between the producer and the receiver. The content must meet the expectations of consumers and the consumers must be able to perceive quality. The quality of content is the foundation of trust for the audiences.

The new media have questioned the type of consumption of traditional media. So far it was enough to have viewers in front of the television set, regardless of whether or not they liked what they saw and whether the content was valuable or prejudicial for their cultural values. The new media paradigm calls for greater participation and involvement of the audience, where their pleasure and satisfaction are key factors in deciding on content development and where the individual knowledge of the tastes, interests and needs of viewers has put marketing over programming tasks. Chapter 7 will focus on these issues.

One of the most pressing issues for the good running of any business is revenue. Current technology allows innovation in the sources of income and a search for more personalized monetization systems. Chapter 8 provides a profound reflection on the economic nature of the goods and services of information and entertainment, and the consequences on its funding and business models – a current debate, unresolved for media companies.

The final chapter is devoted to innovation. It attempts to answer the nature of innovation in media firms. Is all innovation disruptive, or is there collaborative innovation? How should companies innovate? What are the keys to innovation? The observation of some companies will give practical and suitable answers.

I.1. Bibliography

[ARR 06] ARRESE A., "Issues in media product management", in ALBARRAN A., CHAN-OLMSTED S., WIRTH M. (eds), *Handbook of Media Management and Economics*, LEA, New Jersey, 2006.

[DOY 10] DOYLE G., "From television to multi-platform. Less from more or more for less", *Convergence. The International Journal of Research into New Media Technologies*, vol. 16, no. 4, pp. 1–9, 2010.

[EVA 11] EVANS E., *Transmedia Television. Audiences, New Media and Daily Life*, Routledge, UK, 2011.

[MED 15] MEDINA M., *La audiencia en la era digital*, Fragua, Madrid, 2015.

[MUR 11] MURPHY S.C., *How Television Invented New Media*, Rutgers University Press, New Brunswick, 2011.

1

New Paradigms of Audiovisual Industry

The *New World Information and Communication Order*, a concept coined in the 1980s in the UNESCO International Commission for the Study of Communication Problems, has evolved since then. It capitulated in 1990 to give way to the concept of the *Information Society*, as a result of the growth and expansion of information and communication technologies.

Some of the recommendations made in the 1980s are still in force: strengthen the independence and avoid informational imbalances, democratize information and ensure cultural diversity, facilitate access to information, protect journalists, reduce the commercialization of content and expand financial resources, strengthen cultural identity and foster international cooperation.

However, since the 1990s, the world has certainly become more globalized. Against the political powers, citizens have gained greater prominence; therefore, the new international order brings new elements to the industrial ecosystem of media [NIE 08].

The importance of cultural industries as bearers of values and identity led the members of the United Nations to sign the *Convention on the Protection and Promotion of the Diversity of Cultural Expressions* in 2005. The new digital context highlighted the increasingly protagonist role of citizens as content generators, a context where the "creative commons" of the economy of participation and sharing relegated the protectionist approaches of the past concerning intellectual property of works. In the 20th Century, the owners of

---

Chapter written by Mercedes MEDINA.

media left the States in control of the media; in the 21st Century, citizens access content more easily and are even content generators.

In this chapter, we will study how the new order of communication has evolved around the three major issues of concern: the defense of pluralism, the social role of media content and the free circulation of global content.

In short, in this chapter, we will establish the general lines on which the media industry is based; from its global and historical evolution to its triple political, economic and cultural dimension.

1.1. Toward political freedom

Media play a social function that has to do with informing citizens truthfully, showing exemplary models of behavior and providing a useful and effective service to society. These fundamental principles must be secured and safeguarded, because the media shape public opinion and influence the recipients.

When asked who should control the media, political systems offer two extreme positions: interventionism and liberalism. State interventionism believes that the media are state-owned and therefore the public power should control its content and management. The opposite extreme position is liberalism, in which the market itself regulates and directs the activity of the media, and therefore the public sector should not have any external control.

Historically, in most countries, media were directed and controlled by the State, and hence public media were a monopoly. This is the case, for example, of the press in Spain at the time of General Franco or televisions and radios in almost all European countries when they started [CAL 02].

In some countries, totalitarian systems developed systems of censorship, so that only what the government is interested in is published. Although it was a common practice in the past in countries where democracy had not been established, it still exists in some countries. For example, in 2007, the Venezuelan television channel RCTV, which had more than 30 years of experience on the air and was an audience leader, was closed because the government of Hugo Chaves thought it destabilized public order with its content; in China, more than 4,000 officials control the information posted on the Internet or accessed from the Chinese territory every day. When regimes are in a process of being liberalized but freedoms are not fully guaranteed, the censorship system leads to the practice of self-censorship by professionals and media managers. This way, sanctions or other more severe measures against what is published are avoided.

In the 1980s, freedom of trade and expression bloomed in the fields of the media market, in both television and print media. The shadows caused by monopolistic approaches, whose main protagonist was the State, and its dominion over information emerged. In the 21st Century, there are still remnants of statist approaches and lack of freedom in some Arab countries, North Korea, Cuba and China. In other countries, terrorist or destabilizing groups condition freedom of press, as is the case in Colombia, Guatemala or Mexico, where journalists are often killed by drug traffickers [FRE 15].

Liberalization of the media entails the reduction of direct State control, that is, privatizing some public media (as TV channels Rai 2 in Italy in 1975 and TF1 in France in 1987) and the incorporation of private initiatives in the control of the media. When the media were left in the hands of private capital, governments realized the need to regulate the market and establish standards to protect fundamental principles and human rights. For the legal system to organize and allow the full development of the institutions that operate in the free market economy, it is necessary that the law is consistent with the rules of the free market, that it respects free trade, that there is an effective control system and that punitive measures are effective in achieving compliance to the law.

One of the principles that every free state must recognize is that of pluralism. The principle of pluralism recognizes the legitimacy of all cultural options, behaviors, lifestyles, ideologies, policies and human values. The defense of pluralism is justified in considering the public's right to be informed and access the media, and the freedom of expression of journalists. It is essential to guarantee this right and these freedoms in a democratic society.

One way to ensure pluralism is for public institutions to encourage the creation of media companies, so that the different voices of society are represented in the media and have access to them. Thus, pluralism in this sense is identified with a range of media and content wide enough to cover all the information and entertainment needs of citizens.

The liberalization process of radio and terrestrial television in Europe ended in the 1990s. Privatizing the capital and management of media introduced advertising as a revenue source, sometimes the only or main one. Although, historically, the origins of television in North America and Europe are different, we are currently in a stage of coincidence: both in the use of technologies and in the influence that States have in it. Homogeneity is seen in some areas of programming, although the legislation of the Member States of the European Union seeks to maintain a level of European programming. There is another common note between the United States and 'Europe: the States' role gradually ceases to have such prominence and, perhaps above the general interest and public service, commercial forces operating in the market gain strength and influence.

Investment in the launch and implementation of new TV channels, together with limited advertising revenue, required new private partners. In some cases, they came from outside the electronic media industry, for example, from sectors such as banking, computer and electrical. For many of these companies, television was one of the most profitable businesses of their activity. Press managers invested in television, partly to diversify risk and partly in an effort to achieve higher profitability. Most multinational groups participated in television activities: Maxwell, Fininvest, Bertelsmann, Springer Verlag, Hachette, Havas, Time, Warner, Bond and Murdoch. One consequence of the increased presence of televisions in the information market was the stagnation or decrease of the reading of daily newspapers and magazines.

One of the dangers of the free market is corporate concentration, that, ultimately, the media are owned by a few entrepreneurs and, therefore, the information and cultural pluralism is imperiled. To prevent abuse of power, Western legislation established a series of conditions on media ownership, for example, not allowing a single shareholder to control more than a certain proportion of capital – the quantities stipulated often prevent owning more than 25% or 49% of the capital; not allowing one shareholder to own more than one media outlet in the same geographic market, in different markets or any complementary stage of industrialization, for example, that a television channel owns a production company, and preventing one company from overcoming a determined market share – could be applied to both advertising revenue and audience shares. By the beginning of the 21st Century, most Western countries liberalized all these limits to media ownership.

The greater role that civil society has acquired thanks in part to digital technologies has manifested itself in the development of complementary regulatory instruments such as self-regulation. This system leads economic players, social partners, citizens, non-governmental organizations and professional associations to adopt common guidelines and codes of conduct. For this system to be effective, it is necessary that there are professional authorities, independent to political power, to ensure the proper functioning of the market and to resolve disputes out of court.

The existence of councils to help regulate, sanction and encourage free competition and public service is the most appropriate way to achieve pluralism and good practice [LLO 13]. These councils exist in most Western countries. For example: the FCC in United States, Ofcom in the United Kingdom, "*Conseil Superieur de l'Audiovisuel*" in France and the "*Autorità per le Garanzie nelle Comunicazioni*" in Italy. In Spain, the General Audiovisual Communication Law 7/2010 foresaw the existence of a State Audiovisual Council (CEMA), but before it came to exist, the Law 3/2013 of Creation of the National Commission for Markets and Competition decided that the audiovisual sector was to be monitored and controlled by the National Commission for Markets and Competition (CNMC).

Although in the field of media in Spain, codes of conduct on content have not worked, the same cannot be said about the advertising industry, where the Association for Self-regulation of Commercial Communication, "Autocontrol" (http://www.autocontrol.es) is a guarantee of legal compliance and good work [MUÑ 13]. It is a model for many countries in Latin America.

1.2. Media economics challenges

The economic dimension of media activities has been gaining prominence since the 1980s. Thus, media economics has been consolidated as an academic discipline that studies how the media respond to the wishes and needs of information and entertainment of the audience, advertisers and society in general, taking into account the available resources. It has three fundamental elements: (1) the media, which, as a subject of economic activity, is referred to as a company; (2) the product, which is the result of the activity of the company or media and (3) the consumers of the product [HER 09]. These three elements meet in the market, which is known as a physical location or social area where the supply and demand of media products is promoted and met.

The development of the media market in recent years allows us to see how the economic and commercial dimension of communication does not necessarily endanger the cultural, creative and social dimensions of communication. However, we should also be aware of, and reflect on, some concepts that shape the new economic order of the media in the 20th Century.

First, the concept of media product needs to be understood in its double condition, material and immaterial, integrating the concepts of product and service [MED 16]. The media goods concept is associated with a material dimension, the hardware part, while service is associated with the purpose provided. Although the hardware is essential for the information product to be considered an autonomous reality, the value is determined by the content of what is transmitted, the message. The value of the message reflects the kind of needs that media products aim to satisfy. Most goods existing in the market provide consumers with the means to satisfy material needs and desires. News and entertainment products of symbolic and intangible nature can be classified as cultural products. Cultural products deal with the meanings of life, and the needs and desires that they seek to satisfy are directed to the intelligence and will of the individual. Therefore, its value remains and is not destroyed or finished by a first act of consumption.

As we have seen, the guiding role of the State in the launch of radio and television channels in Europe determined the first funding system of audiovisual services, especially television: public funding. The monopoly situation of the first TV channels derived from the concept of public service television and the scarcity of the radio spectrum, owned and managed by the State. Commercial TV channels born from the deregulatory process that took place in Europe in the 1980s and 1990s brought a new way of funding, and consequently, a marketing process. Advertising revenues seemed sufficient to sustain the activity of the few channels that started out in different countries. The first peculiarity derived from this financing system is the presence of a new agent, the advertiser, who pays for the media. The television activity was financed by income received directly from advertisers, but not as consideration for television products. Advertisers do not buy programs, but airtime to insert their ads. Not just any time either, but one that corresponds to programs that bring together the right audience to their interests.

The audience become a commodity when advertisers buy media airtime [NAP 10]. In the case of television, where direct payment has not existed for many years (until the arrival of pay-TV and other systems of video on demand), commercialization of the audience is especially clear. Advertising commercializes audiences, while enabling media to reach out to citizens free of cost. In Spain, the annual advertising investment in media exceeded €10 billion in 2014.

Direct payment for television products introduced price discrimination in an industry hitherto considered free and of universal access. Unlike in daily newspapers and magazines, where direct consideration was customary, direct payment by users of radio and television (and later on, the Internet) took some time to reach Europe. It was not until the birth of HBO (Home Box Office) in 1972 that nationwide industry of pay television was established in the United States. The delay in the development of the pay-television industry was caused mainly by technological difficulties and legal conflicts. In Europe, the first pay-TV channel, Canal Plus France, was born in 1984, modeled on HBO.

Increased competition and a battle for audiences to achieve advertising revenue started a crisis in the public radio and television sector in Europe several years ago. On the one hand, it could be said that the commercialization of media and the increase in media supply made the presence of the State as a guarantor of media service unnecessary. On the other hand, the fact that public media has a fixed income limits its growth potential. In an era of fierce competition in the audiovisual sector, public media, especially in the online world, have been launched into an arena of business operations and partnerships unlike ever before, in order to increase revenue and market presence. This strategy has led to endless debate that criticizes the public service's allocation of money from the citizens into businesses from which they will not benefit. To avoid this controversy, the BBC, according to data from fiscal year 2013/2014, allocated only 5% of the income received from the fee paid by citizens to online activity, while 66% was dedicated to its TV channels, broadcasted in Britain. However, its international activity through *BBC World Service* was funded by advertising and other commercial revenues, which was justified by the need to legitimize their existence and build a global brand, thanks to the accumulation of international audiences.

In short, the dimension of media related to trade and the media economy has gained an especially important role in recent decades. Consideration of audiences as an object of transaction has also influenced the legitimacy of public service activity, which will require high ratings to survive. However, the cultural dimension can also be assessed if one takes the externalities generated by media content into account and the influence they might have on the necessity of public media.

High fixed costs and virtually zero marginal cost (the cost of reaching one more user) are behind the media industry's tendency to operate with economies of scale: especially in audiovisual media and the Internet, there are serious advantages in the size or number of units produced. There are three types of economies of scale:

1) those due to efficient production which therefore can reduce the average costs of production (the higher the number of units produced, the lower the cost per unit);

2) those that take place in the economies of scope, which allow the same product to be reprocessed and used in different ways and

3) those that occur when a product is extensively distributed in the market.

Precisely because the marginal cost is zero and fixed costs are very high, the cost per unit will be reduced not by the number of units produced (as with automobile production, for example), but for the number of people the product reaches. Thus, the first type of economies of scale is not applicable to the audiovisual industry or to the Internet, due to the immateriality of the product.

The second type requires recourse to the definition of economies of scope, which allows the production of two products by the same company to be less expensive for the economy in general than if they were to be produced by two different companies. Economies of scope allow a product to be reprocessed and used in different ways. For example, the same content is distributed in newspaper, radio, television or the Internet and gives rise to differentiated products; or a multimedia group can make effective use of the same content in different media of its ownership. Economies of scope made the fast development of the first digital newspapers a reality (which were originally a copy of the content from papers).

The peculiarities of audiovisual and Internet products give the third type of economy of scale a major role. Because of the high fixed costs, the cost per person decreases with the size of the audience. It is suitable therefore to increase production costs on the broadest possible audience. It can be done, first, with a geographical distribution of the same products in different countries, as Netflix did in October 2015 when it began offering its services in Australia, New Zealand, Japan, Italy, Portugal and Spain.

However, there is also a way of building audiences according to time, something specific to the audiovisual industry and that the Internet has taken further. Content producers intend to distribute their products to the widest possible audience in order to reduce the cost per viewer. Through windowing, for example, they seek to achieve the maximum economic performance of a single product. This strategy involves the transmission of the same program on different channels and at different times. For example, one film is first screened in cinemas; later passed to the home video market; third, it is available on pay-TV channels (also via the Internet) and finally any citizen can enjoy it for free in public or commercial open channels.

This shows that a considerable size is essential if you want to reach mass audiences. In Spain, the large media groups Prisa, Unedisa and Vocento, and later the publishing group Planeta, with its acquisition of a controlling share in Antena 3, were created in the late 20th Century. Consequently, the communications industry has high barriers to entry and a high degree of concentration, in which a few powerful operators set the rules of the game, making it difficult for new competitors to enter. It adopts an oligopoly market structure in which only a few sellers offer similar or identical products.

A logical consequence of the size of the companies is their growth strategy, aimed at controlling all phases of production: content production, packaging and distribution. Vertical integration allows producers, packagers and distributors to become a single agent. The advantages are numerous: access to raw materials, control of the costs of main resources, protection of the marketing channels and differentiation of these channels for better use of the product. This has further developed the Hollywood film industry with giants like Disney-ABC Television Group, which owns the production associated with Disney Studios, and ABC television channels.

Moreover, the growing importance of technology in media has favored the convergence of the telecommunications industries with the media industry. On the one hand, telecommunication companies have seen an opportunity in offering content and, on the other hand, media companies increasingly need telcos to reach their audiences. In Spain, for example, the leading pay-TV in 2016 was "Telefónica", with the acquisition of Canal Plus, which led to additional subscribers of satellite television (with the brand Movistar Plus) and broadband TV.

Part of the growth strategies of companies is the resulting increase in capital needed for investment. Over the years, the industry has gone from being regulated and limiting in property matters to prevent dominant shareholders in the interests of pluralism, to a jump to the stock market in search of financial resources. This evolution implies the presence of shareholders, who want dividends and value the equity returns in the stock market. The audience was seen as an object of trade with advertisers, shareholders and media companies taken into the logic of the stock market: a logic run by the pursuit of profitability and short-term risk. This explains why issues such as the quality of content, which is achieved in the long term, are not among the priorities of the companies. The stock market is usually quite unpredictable, as many factors influence its movements. A sector may be booming, but if an influential factor arises unexpectedly, the value of the stock can completely fall. In Spain, in 2016, only one media company, Mediaset, was among the 35 listed companies with the highest liquidity on the Spanish stock market (IBEX 35).

However, with the creation of large conglomerates and their presence in the stock market, we have witnessed in recent years the development of content and communication platforms that exceed the constraints of oligopolistic structures. As a means of distribution, the Internet is a network of networks with great flexibility and consequent impact on the cost structure. Furthermore, the development of any project is not related to a specific geographical territory, and the structure of *the network of networks* makes it universal and accessible from practically anywhere.

The geographic and legal barriers (spectrum designation, concessions, etc.) are non-existent; only language impedes the content from traveling anywhere on the globe. This has made it possible for citizens to create their own content and disseminate it. Furthermore, this circulation of media content greatly depends on the active participation of citizens. In this sense, the convergence brought by the Internet does not refer only to a technological process, but to a cultural change. We are thus faced with a market where oligopolistic structures and the consolidation of large conglomerates are compatible with the potential of end users creating and distributing their content without limits.

The high dependence on advertisers for revenue collection gave the media a great financial instability, as the ratings, the sale of copies, visits on the Web, etc. conditioned the amount of income. The collapse of advertising revenue caused partly by the loss of readers and audience segmentation, exacerbated by the financial crisis that hit Europe and the United States since 2007, caused a willingness to produce cheap content that could reach large audiences. Quality content was replaced by profitable *commodity* content, designed to entertain more than to inform with rigor.

Indisputable success formulas like reality shows, contests, tabloids and sensationalistic headlines stopped investigative journalism, high-quality productions and the creation of relevant stories [MED 13]. In addition, the speed of the digital environment permitted professionals to disseminate news without confirming their sources or to publish content without prior review.

The risks of the media market from the 1980s evidently did not dissipate over the years, but grew with some of the new offers and modes of communication.

Nevertheless, entertainment content and content compelling to large audiences are not an obstacle *per se* to the pursuit of quality that the public demands, and for which it is willing to pay, if necessary. The popular programming for the masses has legitimated the activity of TV channels such as the BBC, where a brand value associated with high-quality standards has matured; or some drama series like *24* or *Lost*, seen by audiences worldwide, or shows like *Pop Idol* or *MasterChef* of great success among national audiences.

1.3. International trade of ideas

Because of its cultural character, information products have the ability to promote knowledge, experience and imagination of individuals. In economic terms, it seems that information products have externalities, understood as the influence of a person's actions on the welfare of another. If the influence is negative, it is called a negative externality; if positive, positive externality. Therefore, information products can be classified as desirable goods (merit goods), just as health, education and national defense, because of the positive externalities that they are likely to produce. The opposite of a merit good, one with negative externalities, is discouraged by society in general. However, the cultural character of information products makes them undesirable in some cases and, as discussed in the previous section, the State or other mechanisms of self-censorship from the profession itself may intervene. Noteworthy is the concern of parents and authorities over the impact of content or video games on children and young audiences.

As noted above, in the early stages of television and especially since the 1950s, shows too have become objects of sale and begun to circulate around the world. It is not enough to consider the exchange of shows as if they were material goods; it seems necessary to pay attention to the dissemination of ideas to other countries. For some, the problem lies not only in the dissemination of foreign ideas but also in the destruction of local cultural values. In academia, the critical school that stems from Marxist philosophy coins the term "cultural imperialism". It emphasizes the danger that the invasion of US programs represents to the world. The first author who applied this expression was Schiller in 1976, followed by Said, Mattelart, Dorfman, Varis, among others. Research by Varis [VAR 84] shows the power of the Hollywood industry as a threat, as the only center of production and export to the rest of the world.

However, for others, of the liberal school, telecasting of American TV shows in other countries is a consequence of the free market economy, and it is as an opportunity for the domestic industry to grow. In particular, Faus [FAU 95] points out that as a result of their extensive experience in the production of television programs, a standard narrative and visual levels of understanding has been achieved, hardly attainable by the average production of any other nation. Moreover, as Bertrand [BER 86] points out, the imperialist vision doubts that the consumer decides how to make use of the material supplied. According to his opinion, one who does not want to be dominated will not be dominated.

Moreover, it is sometimes ignored that there has been a strong flow of shows exchanged between America and Europe, sometimes with help from national governments who established agreements with other countries to channel the buying and selling of televisual products, and also between Latin

America and Spain, or even the great export of Japanese cartoons. At present, there are large producers of content for the international market. For example, Brazil, Venezuela, Argentina, Mexico and Colombia produce soap operas that are watched all around the world, and India and its Bollywood film industry or the fledgling film production in Nigeria.

Nonetheless, we must recognize that this international exchange of programs has led to the homogenization of the content. The new countries that were incorporating television to their media market had the pioneering countries as models for the production of programs, especially the United States and Britain. We could say, therefore, that a few countries provided to many, and that in each country the sector was in few hands. This situation suggests that the new international order of the 1990s revolved around the Western world and in particular English-speaking countries, and that in the new digital scenario this is happening again, since the vast majority of Internet domains comes from the United States and the United Kingdom. However, this does not prevent each country from having developed an audiovisual industry around its local television.

The European Union wanted to protect their culture from the threat of the United States and established quotas of European programming and independent production, which has become a commercial protection more than a guard of cultural identities.

European audiovisual policy intended to achieve a closer union among the European peoples through cultural unity. The policies considered television an essential instrument to promote these cultural purposes. These cultural reminiscences regarded television as an essential public service, more related to earlier times than to the reality of the 21st Century.

In particular, this policy was orchestrated in the *Television without Frontiers* (TVWF), which was amended several times. The latest version is the European Directive 2007/65, *Audiovisual Media Services Without Frontiers* (AVMSD). European Directives propose minimum measures to be taken in each member country. In order to boost European production, Article 4, updated by the AVMSD, states that broadcasters should allocate a majority proportion of the transmission time (excluding the time appointed to certain formats such as news, sports events or advertising) to European works. This proportion should be accomplished progressively, and must in any case not be lower than the average of 1988 or 1990 in the case of Greece and Portugal.

The most important improvement of the revisions was to include the definition of “European work”: a work made essentially with authors and workers residing in one or more Member States of the European Union. Furthermore, it has to meet one of the following three conditions: that the works are carried out, supervised and effectively controlled by producers established in those States and that the contribution from the co-producers of those States in the total cost is a majority.

Accordingly, Article 5 defines that broadcasters must reserve at least 10% of their transmission time (also excluding the time appointed to certain categories as in Article 4) or at least 10% of their programming budget for European works created by producers who are independent of broadcasters, half of which must have been produced in the last 5 years. In Spain, the General Law on Audiovisual Communication (LGCA) 7/2010 substantially clarifies the legal definition of the *independent* producer. Given that the producer is the natural or legal person who takes the initiative, coordination and economic risk of the production of audiovisual content, the independent producer is the natural or legal person producing such content on its own initiative or on request and, in return for consideration, allows a provider of audiovisual services that is not linked stably with it in a common business strategy to air it.

The development of content distribution systems has made it possible for information and entertainment products to be enjoyed by the user beyond the control of content creators and/or owners of rights.

The contents scheduled on the programming grids in the corresponding commercial time strips are available on different websites, freed from the constraints of time, and therefore outside the conventional commercial possibilities. In some cases, this is part of a bigger strategy of multiplatform content development, as content is distributed through official sites including commercial messages. In other cases, they are available on unofficial sites for fans. This reality raises questions of authorship and intellectual property. However, in recent years, it is also developing the reverse process, and from the extension of content to the Internet, we are seeing the creation of content by end users.

The evolution of Web 2.0 lets us formulate a demanding concept of what we mean by user-generated content (UGC). As a concept, the content generated by the user can be defined by three fundamental elements:

– first, UGC is content that is posted without barriers or restrictions but is not produced by the user for interpersonal communication, like an e-mail;

– second, it requires some creative effort, either for the development of new content or adapting existing content;

– finally, the creation of this content remains outside professional practices and routines.

Regarding the types of UGC, we can distinguish mainly between blogs, wikis, podcasts and social networks. Applications for mobile phones and tablets could also be included in this classification.

The UGC also receive the name of consumer-generated media (CGM). This definition extends the generic relation of content to media, more inclusive in its role of agents of the media market.

Among the characteristics of user-generated content, we would emphasize first the great sense of ownership of the media that gives rise to this content. UGC helps to cultivate relations between common cultural interests, and along with it, facilitates social integration. As a result of these relationships, recognition arises for some amateur artists who are discovered by agencies, websites and media companies.

Along with these realities, UGC blurs the difference between editor and user, with the problems of authorship and attribution that entails, especially in the information field. A result is that piracy seems easier as the perception of illegality and crime is minimized. Moreover, it is common that the content is of low quality, although some may be considered as a seed of creative talent. Some creations have made the leap from the Internet to the conventional grid of national broadcaster, like the series "Qué vida más triste", born for the Internet and then aired by the television channel "La Sexta". The network considered it a product that fits very well with young audiences and their line of unconventional programming. It began its broadcast in October 2008 and reached 1 million viewers.

Regarding the advertising possibilities of this content and throughout the digital context in general, we are still at an early stage. Traditionally, advertisers have bought advertising time or space in conventional media, in a controlled context: they knew where the ad would appear, how it would be and in what environment it would be seen. Considerable success for a prolonged time is required for user-generated content to attract advertisers.

Its accessibility, together with the reduced production costs and zero marginal cost, boost the Internet as the ideal platform for distributing content created by users. The very nature of this network gives creativity very few barriers. Until the advent of the Internet, for content to reach large audiences, or even start production, it had to go through a very long and competitive process, in which commercial criteria often prevented its successful development. Many of these processes are a synonym for quality control: checking the veracity of the sources, the scripts, the creation of characters, etc.

Precisely because of the absence of hindrances to creativity, much of the fictional UGC places special emphasis on transgression, irreverent and erotic content, which would have difficulties being shown in traditional media and somehow responds to desires of protest, social integration, etc. Therefore, quite often, the lack of artistic and technical quality is one of the most negative notes of user-generated content. For this reason, the productions that could qualify as hotbeds of creative talent are rare. It is undeniable that the possibilities for true artistic talent to be developed through the Internet are unlimited.

1.4. Bibliography

[BER 86] BERTRAND C.J., "Imperialismo cultural", *Nuestro Tiempo*, vol. 390, pp. 110–123, 1986.

[CAL 02] CAL R., "Las presiones del poder sobre los medios de comunicación social", *Historia y Comunicación Social*, vol. 7, pp. 11–30, 2002.

[FAU 95] FAUS A., *La era audiovisual. Historia de los primeros cien años de la radio y la televisión*, Ediciones Internacionales Universitarias, Barcelona, 1995.

[FRE 15] FREEDOM HOUSE, Freedom in the World, available at: https://freedomhouse.org/report/freedom-world/freedom-world-2015, 2015.

[HER 09] HERRERO M., "La economía del producto audiovisual en el mercado de la comunicación", *Comunicación y Sociedad*, vol. 22, no. 1, pp. 7–31, 2009.

[LLO 13] LLORENS C., COSTACHE A., "La gobernanza como marco teórico para el estudio del pluralismo de medios", *Trípodos*, vol. 32, pp. 63–76, 2013.

[MED 13] MEDINA M., "Contenidos de los medios en el entorno digital: el poder de entretener", in GUTIÉRREZ R. (ed.), *Poéticas de la Persona. Creación, responsabilidad y vigencia en la Comunicación Pública y la Cultura*, Comunicación Social, Salamanca, pp. 64–72, 2013.

[MED 16] MEDINA M., ARESE A., SÁNCHEZ-TABERNERO A., "Content as products in media markets", in LOWE G.F., BROWN C. (eds), *Managing Media Firms and Industries. What's So Special About Media Management?*, Springer-Verlag, pp. 243–261, 2016.

[MUÑ 13] MUÑOZ M., GÓMEZ-IGLESIAS V., "El protagonismo de la autorregulación y de la corregulación en el nuevo mercado audiovisual digital", *Trípodos*, vol. 32, pp. 77–92, 2013.

[NAP 10] NAPOLI P.M., *Audience Evolution: New Technologies and the Transformation of the Media Audiences*, Columbia University Press, New York, 2010.

[NIE 08] NIETO A, "Ciudadano y Mercado de la Comunicación", *Comunicación y Sociedad*, vol. 21, no. 2, pp. 7–33, 2008.

[VAR 84] VARIS T., "The international flow of television programs", *Journal of Communication,* vol. 34, no. 1, pp. 143–152, 1984.

2

The Threat of OTT for the Pay-TV Market

The television industry is relatively new and therefore, naturally open to innovation and highly influenced by technology and the consequent improvements in the quality of delivery or consumption.

High entry barriers, strong monopolies or oligopolies and big multimedia conglomerates in control of as many stages as possible of the value chain have also characterized this industry. It is in this context that the disruption of OTT services shook the traditional business models of legacy media and forced them to face more tech-oriented new players, capable of satisfying more or less the same needs and wishes in apparently more efficient ways.

With the recent global expansion of the famous video service, Netflix, the question of whether Netflix will succeed in making its new host countries a profitable home or not has been raised worldwide. The service arrived in Spain amid great anticipation and uncertainty. Its pay-TV operators and traditional broadcasters had been experimenting with video-on-demand (VOD) streaming services and struggling with revenue models in an audiovisual market undergoing constant restructuring and with a low pay-TV penetration [MED 15a].

Looking at it either from the side of the national broadcasters and its developing competitors, or from the side of the new entrant, the current situation

---

Chapter written by Alicia URGELLES.

is very interesting to look into: how are the national industries working so far and how much space is there for the American giant to take in the ecosystem, if any? How will Netflix deploy its forces in the new territory and what strategies will it follow in order to conquer the new markets, if it can?

In this chapter, we aim for a better understanding of what it means for a national industry to face the entry of this new international competitor with such and innovative and successful profiles in terms of shaping consumption culture, while looking at the competitive landscape in which content creators, aggregators and distributors operate.

2.1. The "new" barriers to entry and value chain

Over-The-Top (OTT) services deliver content over the Internet, "over the top" of traditional infrastructures, set-top boxes and traditional cable or satellite pay-TV. They allow users to sidestep incumbent pay-TV operators, with their high fees and limited bundles, for a connected and on-demand television experience. Broadband connectivity provides instant access to a global network of services and applications, "enabling equipment manufacturers, application providers and customers to bypass the services offered by traditional network operators and to cross national boundaries, thus obtaining multiple new routes to market" [GOD 15, p. 11]. OTT services could substitute, to some degree, traditional media and telecom services, so pay-TV providers fear for the loss of their lucrative business, while online services experiment and sometimes succeed in discovering revenue models. The panorama of OTT SVOD (subscription-based video on demand), AVOD (ad-supported) and TVOD (transactional) services is growing.

According to Porter's theory of the five forces, the threat of new entrants shapes the competitive structure of an industry. The threat of new entrants refers to the threat new competitors pose to existing competitors in an industry. "New entrants to an industry bring new capacity and a desire to gain market share that puts pressure in prices, costs and the rate of investment necessary to compete" [POR 08].

Several factors determine the degree of the threat of new entrants to an industry. Many of these factors fall into the category of barriers to entry. These are factors or conditions in the competitive environment of an industry that make it difficult for new businesses to begin operating in that market. The high fixed cost and almost zero marginal cost of the audiovisual industry have served as a traditional entry barrier in the industry. These costs are not only related to the production of content but also to the infrastructures that make broadcasting, cable and satellite television work, making legal and technical requirements another important barrier.

Given the flexibility and universality of the Internet as a means of distribution, as well as the absence of major geographical and legal barriers, audiovisual products distributed through the Internet are ideal to develop economies of scale and windowing strategies in a more efficient way than ever before [HER 03]. With the digital economy, traditional entry barriers in the audiovisual industry have been mostly diffused to the extent than even final users are able to produce and distribute audiovisual content. Therefore, it seems like new entrants may find different entry barriers.

These changes place content itself in the center of the television business. Content, along with the viewing experience of the consumer, becomes more important in shaping of the strategy. There is a daily battle between traditional content companies, distributors and technology companies for control of the viewing experience. The viewing experience itself has also changed: with the definition of television having changed so much that it is not limited to the device anymore, companies that provide services for a viewing experience have been forced to face as competitor players who were not their natural rivals.

We could easily assert that real entry barriers have to do with the ability to offer relevant content and reach the audience in the most convenient way to them. "The Internet tends to dampen the bargaining power of channels by providing companies with new, more direct avenues to customers" [POR 01, p. 66]. In a world of limitless choice, where almost anyone can acquire, create and distribute interesting content, the winner will be the one that can utilize data to respond to audience demands most nimbly and drive an experience that feels personalized.

Mapping the structure of the television industry has become confusing and not as straightforward as before. In this context, there is also the threat that new entrants become substitutes in the audiovisual industry, and some prominent providers disappear. According to Ala-Fossi and Lax [ALA 16, p. 1], “a number of recent reports suggest the life expectancy of broadcasting as a free-to-air television platform may be no more than 15 years”.

The concept of the value chain popularized by Porter has helped us view the processes that make up the system behind the audiovisual entertainment market and the activities that need to be carried out to create competitive advantage. Nowadays, and with the influx of the digital economy, an important topic in media economics regards the transformation of this value chain. Times, players and windows for creation and supply of contents are changing; shorter times require faster adaptation to novelties, more experimentation and higher concentration in legacy media competing with a flexible smaller player.

The “sidestepping” that comes with OTT services in the audiovisual world could be described somehow with a comparison to *Uber* disrupting the supply chain of taxis: large taxi companies that manage their fleet of taxi drivers are sidestepped when another player allows a seeming direct contact between the drivers and the customers. Intermediation still exists, as the app does the work of connecting drivers and users, but to the user, everything is more direct, faster and more efficient.

The perspective of the users is very important. Companies tend to forget that their customers do not see the intricacies of the value chain or the supply chain when they make their purchases. Viewers tend to be unaware of the processes and activities that led to the content they are accessing, and who manages logistics, operations, infrastructure or marketing. They relate to the point of access to content and the quality provided by that immediate service, a quality that corresponds with what they have paid and, therefore, with that value delivered in the chain. That is why it is necessary to define services in terms of how they are viewed by consumers rather than according to the technological or managerial elements involved. Viewers want content, and the easiest and cheapest way to access it will go a long way in the market.

An example of this redefinition of the value chain in the audiovisual world can be seen in the case of *Spotify*. The digitization of music ruptured the music value chain: an industry that was used in strong margins did not react fast enough to digitization and its culture of gratuity and piracy. Amid the crisis, Apple's iTunes opened a window for those who would still want to pay for music outside of the limitations of albums and physical copies. With *Spotify*, the rules of the music industry changed further. A technology-driven business model replaced traditional intermediaries while answering to an idea of "access" over "ownership". *Spotify* exploits the fact that a large number of people nowadays do not want to pay for specific music products to own them in the form of a physical library; many prefer to subscribe or gain access to a "database" they can navigate, control and share. Any streaming model these days follows this desire for control and vastness that characterizes modern consumers.

Real disintermediation does not happen, but, with the new intermediaries, distribution costs are lower and the chances of smaller creators reaching end users increase. Something similar can be seen in the case of the famous *YouTubers* – users turned content creators turned celebrities – who reach massive audiences through their online channels and vlogs in the video service *YouTube*, without the complicated system of processes and activities, that is, television broadcasting.

All these online services have another entity in their favor that traditional broadcasters do not: the real-time feel of their audiences (comments, "likes") and the colossal amount of data they can gather that can help them improve and react faster to their demands. If well managed, the information can be used as a commodity – in the same way, for example, as in the traditional advertising-funded TV model – or as the research basis needed to start new complementary businesses. Geo-marketing, geo-location, geo-targeting and similar activities are taking advantage of these data possibilities.

OTT has also been an enhancement for mobile consumption. Research on the role of different devices in media consumption has become popular recently and the understanding has grown over the type of device preferred to watch different content (length, platform, genre, quality, etc.). Mobility is yet another feature than can be gained with the "sidestepping" that comes with OTT. In the case of Netflix, for example, the IHS Technology Connected Device Market Monitor report showed that as of the end of 2015, Netflix addressed 339 million connected devices in the United States, securing a presence across 32% of the connected devices in the country. According to a Nielsen report, 47% of Spaniards now consume VOD content, which is approximately one in two Spanish people [FER 16].

Telefónica, the Spanish telecommunications giant, is going through an important expansion phase of its video services in Latin America, launching their OTT video platform. The special combination of technology and content that leads the OTT business is perfect for the media's economies of scale; once the technology is developed and tested successful, and the contents are produced and tested, it is just a matter of sealing business deals and expanding geographically, which is what Netflix has done.

2.2. The pay-TV competitors in Spain

In Spain, there are three main types of TV providers: broadcasters, telephone carriers (Cable, ADSL) and streaming service providers. The competition in the pay-TV sector has grown recently; furthermore, a process of concentration has taken place. There are many competitors with a certain grade of differentiation. Recently, there has also been an increase in competition.

One of the oldest and the biggest competitor is *Canal+*. It was launched in 1990 as a pay-TV channel in a monopoly regime. After different transformations toward a multiplatform digital television via DBS, in October 2011 it launched a service called *Yomvi* to spread its offer to the Internet, allowing subscribers to access content from every device and at any time. In June 2014, *Telefónica* bought 56% of its shares. Therefore, *Telefónica* controls 78% of Digital + and since then its joint bit was called *Yomvi Movistar+*, using DBS and ADSL transmission systems.

The cable industry as a pay-TV competitor started in 1996 after some difficulties to get the license to operate. After some mergers, the leader has been *Ono* for many years, which operates in most of the regions of Spain. *Ono* offered TiVo as an experience of Smart TV, since it allows enjoying apps and Internet content from the television set thanks to a special fiber-optic cable for the service. Although there are three more providers in the market – *R*, *Telecable* and *Euskaltel* – they do not compete among themselves because they are regional providers in specific regions. Moreover, *Euskaltel* bought *R* in 2015. In 2012, Ono's debt was €3.44 billion and it was bought by *Vodafone* for €7.2 billion. The current leader of the pay-TV market is *Movistar* (with 3.9 million subscribers in 2015), because of the merger with Canal+. The next competitor is *Vodafone* and thanks to the acquisition of *Ono*, the number of subscribers in 2015 was 876,000.

The main differences between satellite and cable were price and added services. For audiovisual content, cable operators offer triple and quadruple play (the Internet and telephone, both fixed and mobile). The acquisition of *Canal+* by *Movistar* has added to *Canal+*'s content, the added value of the Internet and telephone line. However, the advantage of this company against the rest of the new OTT and VOD services is the supply of sports, especially soccer matches.

The owner of soccer, basketball and Formula1 rights is *Mediapro,* which has 50% of the DTT channel *beIN Sports* – the other 50% belongs to Al Jazeera. *Mediapro* offers the matches of the Champions, Europe and the Spanish League through its DTT channel *beIN Sports,* the OTT *Total Channel*, the telecommunication companies *Movistar*, *Vodafone*, *Orange* and *Telecable* and *YouTube* since 2015. Soccer PPV was a very successful model from 2002 to 2007. However, from 2008, it lacked business chance and was mainly substituted by subscription contract and monthly fee model.

Although the number of movies available since 2009 was higher than soccer VOD, the demand of PPV movies decreased since 2006. This is a strange trend because since 2006 new online services (OTT) have been launched in the Spanish market: *TotalChannel*, *Cineclick* and *ADNstream*. Some of them belong to traditional distribution companies such as *Filmin* or *Filmotech*. *Filmin* is specialized in Spanish movies, and it distributes independent films. Others like *Pop Corn,* *Wuaki TV* and *Netflix* are international companies which started offering premium content for online services, especially movies and television series. *Wuaki TV* was a Spanish start-up that was bought in 2012 by Rakuten, a big Japanese e-commerce platform.

The main business models of these companies are PPV and VOD, but there are also subscription-based contracts. Most of them are streaming, but there is also the chance to download content. What happens with some of Spanish OTT companies is that they might close down after 1 or 2 years, because the competition is so high and the level of piracy in Spain is high. Furthermore, the content is not good and the bandwidth is not sufficient. According to the survey run by consultant Cocktail Analysis, only 26% of Internet users accessed pay online video in 2013 [COC 15]. On the contrary, according to the same source, people who have contracted pay-TV consume mostly football and NBA matches, and they used *Yomvi Movistar+* and *Wuaki* TV.

Data from the market that most of these OTT services are not well known by the general public and they do not use them to watch videos: only YouTube and Google Play are well known as video services in a payment base. The rest of them are less popular for the general audience [COC 15]. Traditional broadcasters offer premium content through their online services: The *Atresplayer* catalogue is broader than *mitele* from Tele5 that only offers movies on a PPV basis [MED 15b]. *Nubeox* was part of *Atresmedia* group, but since 2016, it has been integrated as *Atresplayer Premium*, the premium app of the broadcaster *Atresmedia*. The hardware producers, such as *Apple TV* or the e-commerce platform *Amazon*, supply premium audiovisual content as well, but they are not the most used by the general population [FAH 16].

Nowadays what all these providers have in common is that content can be watched on every device: Smart TV, Smart phone, tablet, computer, on Chromecast or PlayStation. Regarding the differentiation in the type of content delivery, the distributor system goes from DBS (*Yomvi*); cable (*R* and *Telecable*); ADSL telecommunication companies such as *Movistar*, *Vodafone* and *Orange* and the OTT operators. The number of subscribers makes a difference in the bargaining power of customers, because some systems would have gotten a higher penetration than others. Figure 2.1 shows the number of subscribers for each distribution system.

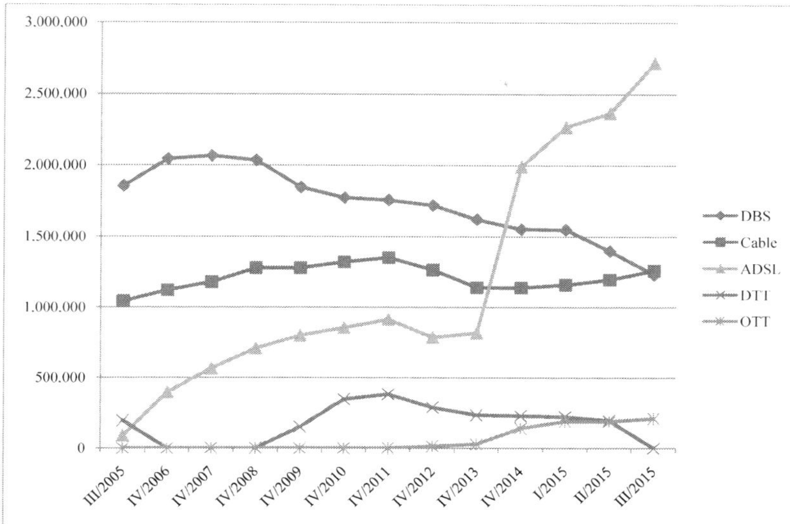

Figure 2.1. *Number of subscribers by distribution system (Source: CNMC (03/08/2016)). For a color version of this figure, see www.iste.co.uk/medina/audiovisual.zip*

This is the competitive scenario where Netflix interrupted in October 2015. The next chart will summarize the current supply of pay-TV in Spain to understand the competitive environment that Netflix needs to overcome.

DBS/ADSL	Cable	ADSL	IPTV	OTT	DTT
Movistar +	Ono Euskaltel/R Telecable	Orange Vodafone	Atresmedia Mitele	Filmin Filmotech Wuaki AdnStream	Axn BeinSport
80€/month	82€/month	97€/mnoth	1-3€ ppv	7-10€/month	10€ ppv
Sports Soccer Docus Premier movies TV Series **Internet** **Telephone**	Sports Soccer Docus Movies TV Series 70 TV channels **Internet** **Telephone**	Movies TV series Soccer Wuaki CANAL+ Series **Internet** **Telephone**	Movies TV series TV shows	Spanish movies Old movies Kids Movies TV series Docus Cook Current affairs **Blog**	Movies Sports

Figure 2.2. *Spanish supply of pay-TV (distribution system/companies/ prices/content and services) (Source: own elaboration)*

To access pay-TV, Spaniards have different technological distribution systems; apart from the technological standards, every system offers different content and services. Only ADSL and cable are able to offer the Internet and telephone at home, which is why the monthly fee is very expensive. The apps of broadcasters work mostly as experimental tools to offer free TV content for around 3 weeks before they turn premium. As it was said before, the OTT companies offer some commercial movies and two of them, *Filmin* and *Filmotech*, are the market place for Spanish movies. Sports are one of the most attractive contents for payment, and it is possible to access them on *Movistar+*, the DTT *BeInSport* and *YouTube*.

Despite the low general awareness of OTT platforms and online video providers, the Spanish market had been expecting the arrival of such a renowned platform as Netflix, and expectation had to be managed about the values and limitations of the service. The three key items for which the platform has come to be known are its expanding catalogue – especially its "original" content which have gained the favor of critics and subscribers – [DOU 15], the affordable, no-commitment monthly fee that allows for unlimited viewing on any Internet-connected screen [NET 16b] and, finally, their highly personalized user interface and recommendation engine. How capable would Netflix be of offering these brand staples in Spain? If or when it does, how willing would be the local market to pay for the said offered service?

2.3. The competitive and cooperative strategy of Netflix

To analyze Netflix, Small [SMA 12] proposes three key items of its value proposition: a large catalogue of content for users to choose from, a useful and innovative recommendation engine and numerous streaming platforms to access. Vonderau [VON 15] describes the complexity of video streaming pricing discrimination strategies: minimum guarantees combined with revenue sharing flat deals (i.e. one-time advance payments), package deals (covering batches of titles) or output deals (monthly compensation). Each of these patterns comes together with one specific type of exploitation rights. Jenner [JEN 16] questions the advantages and disadvantages of the added value of Netflix summarized in "to watch instantly". The narrative style of the television series demands seasonality and suspense from the consuming experience, online streaming and unlimited catalogue of the seasons of television drama.

Apart from Netflix's own strategy, it is particularly interesting to observe the actions of the traditional players in response to the developing content environment and changing consumption culture [MED 15b]. In this context, suppliers are especially important.

Although it is still early to make conclusions about the platform's overall success, this chapter suggests that the conditions and market structure that Netflix finds in Spain will result in limited success of the company. The starting point to launch Netflix successfully in the United States is completely different from that of Spain. Among other factors, the illegal downloading practice is quite extended, and the competition in VOD and other services is high.

Around the day of arrival of Netflix in Spain, the service was also launched in Portugal and Italy. As can be seen in the next image (Figure 2.3), a few months later, on January 6, 2016, the service reached another 130 countries, and Netflix could officially call itself "global".

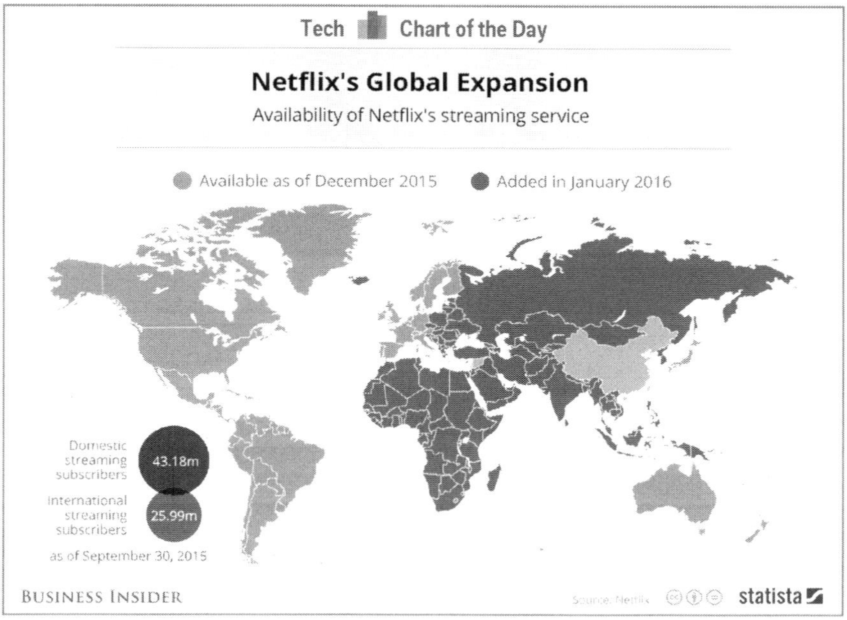

Figure 2.3. *Netflix International Expansion (Source: Business Insider). For a color version of this figure, see www.iste.co.uk/medina/audiovisual.zip*

What can be observed about the company's specific strategy in Spain stands only as a part of a larger strategy, for a bigger international panorama. Much has been said about the worldwide expansion of the company and strategy behind its drive for international subscribers. In the Spanish case, and taking into account the depicted pay-TV market, it is possible to look into what may look challenging for the permanence of service.

From the viewpoint of consumers, the biggest challenge to overcome may lie with content. Content is a key factor in the strategies of players, to attract consumers and differentiate their offer, and, in Spanish pay-TV, the market of content suppliers has been highly concentrated. As we have stated before, the new digitalized television ecosystem has lowered entry barriers and left the battle for engaging content as the most obvious one, especially for pay-TV.

The building of Netflix's video library demands the negotiation for the territory of three types of content: (1) those series and films commissioned by Netflix and known as "Netflix Originals"; (2) the third-party content, licensed from foreign studios or networks in order to fill the catalogue, and (3) the content that the company can license or commission from national sources. Suppliers and their bargaining power can have a significant impact over a company and its profitability, which is why a big part of the strategy of Netflix in Spain resides with its suppliers of both content and infrastructure.

Netflix defines itself as an "Internet TV network" [NET 16b] and it works as such. Not a studio but a distributor, its latest strategy to focus more on ordering original production and less on licensing existing shows suggests a very similar way of working to traditional television networks, applied to the Internet context. Strategically, management of the rights for its content is the key issue [DOY 16].

For example, in Spain, the rights to Netflix's most famous series, "House of Cards" and "Orange Is The New Black", are in the hands of *Movistar+*, as well as exclusive HBO production for which users seem willing to pay. However, the list of originals grows steadily, with appealing announcements of future projects, and is becoming an important attraction for the networks' target of series' fans [GON 15].

Netflix's arrival to a new country means not only a source of new subscribers but also a cradle of new international content. As chief content officer, Ted Sarandos, explained during the launch event in Madrid in 2015, the service is "looking for great storytellers and worlds that people have not seen before and that can travel around the world – stories that are entertaining in Spain, but also for our viewers around the world" [ROL 15b]. They have recently announced that a new original series to premiere in 2017 will start production in Spain in 2016 [NET 16a], which will strengthen the relationship the company already has with Spanish producers like *Bambú Producciones*, and give way to other important alliances with national industry. Netflix has already licensed some successful series from broadcaster *Atresmedia*, like "Velvet", "Gran Hotel" or "The Time in Between", as well as films co-produced by *Atresmedia Cine*.

This movement into the production market might win over Spanish content creators; however, it does not necessarily translate into more subscribers in this market. National open broadcasters are strong on local fiction and the content that drives most of the aforementioned pay-TV scenario are international and sports. Time will tell to what extent the service will be able to bid for rights for global distribution of attractive third-party content before its Spanish competitors, as well as to what extent will it need to. Netflix's costs of content have risen as the service expands internationally. Content costs will only increase when its rivals join forces to make more competitive bids, and Netflix still needs to transform those investments in international content into profits and maintain their growth in subscriptions.

The second key value that consumers link to the brand is "the affordable, no-commitment monthly fee" that "allows for unlimited viewing on any Internet-connected screen" [NET 16b]. Price is an important issue in the Spanish context, where open television and high levels of piracy have enforced a culture of gratuitousness concerning audiovisual content [ROL 15a, ANT 15].

Netflix in Spain offers three different price plans for its subscription service ranging from €7.99 to €11.99 a month – the first month for free for all new subscribers. Its price system relies on additional features – higher-definition quality and the amount of screens in which it is possible to stream at the same time – for price discrimination, not on the quantity of content available. This reinforces the sense of abundance related to the service, while reducing the complexity and frustration that other types of packages and plans based on catalogues presented to users. The prices remain in line with the ones in other European countries, which also might reinforce the perception of the brand's international unity and potentiality.

The library of content in Spain may not be as vast as the one in the United States, but it is still large, expanding and working toward being a worldwide company, or as CEO Reed Hastings assured: "It's a temporary issue. We know what everybody wants, which is the same catalogue around the world. We're frustrated that we're not there, but we will get there, and before everybody else" [BAR 16]. Along with the portability of the service and no interruptions for advertising, the possibility of creating up to five profiles with one account makes paying for content seem a probable alternative for Spanish users.

The third differentiating value element Netflix offers is its personalized interface with data-driven recommendation engine. Freedom from the grid gives more options to VOD users, yet what really distinguishes a warehouse of un-ordered content from a personalized library is the help the user gets in picking interesting and satisfying options among the abundant possibilities. Users are placed in a position that combines the power to choose with the satisfaction of more certain choices and an overall more personal experience. This system benefits the satisfied viewer as well as the company, which is able to "create and stimulate demand for content" [SMA 12, p. 52].

Moreover, the fact that Netflix "does not have to justify itself to advertisers" [JEN 16, p. 262] – being only accountable to shareholders and its partners in revenue-sharing – might help make the company much more "user-friendly" from the viewpoint of the Spanish consumer, who endures television with many minutes of advertisement per hour [OBS 15]. According to a report by Nielsen on VOD consumption in Spain, the audience of online platforms see advertising as an annoyance. Over 65% said they would like to be able to block advertising, as they do not identify with the advertised products [FER 16].

The measurements for success of series and movies in linear advertising-funded TV do not apply to Netflix. The network orders entire seasons of series and presents them to its viewers for binge-watching, breaking traditional seasonality timing: times of production and times of consumption. The new times and stipulations for content commissioning and bids for right give Netflix an advantage over smaller national players, who may be stronger in local battles but are worse endowed for an international war.

In a context with high levels of piracy like Spain, where introducing the culture of paid content to newer generations seems important, the consolidation of Netflix and similar SVOD are special achievements. As stated by Vonderau [VON 15, p. 729], at least, "streaming revives the belief in the magic of music or video as commodities". However, it is more than that with Netflix, because it is not simply a platform for video as a commodity but a space for video as a service.

Part of what makes it difficult (and early) to measure the success of Netflix's arrival in the Spanish television market is the definition of "success". The current lack of available data from Netflix does not help, and there are no certain answers yet when comparing and contrasting thresholds and figures of revenue, subscribers or profits. In business terms, success means to obtain extraordinary results to overcome the expenses and produce benefits for investors. In most of the countries, except for Mexico and Colombia, Netflix seems to have interrupted as a small competitor, not behaving as a real threat for the rest of the competitors. It is clear that Netflix's strategy in this primary stage is getting a bigger market share rather than profits: a land-grab strategy: catch attention and market share before it is too late and figure out revenue and business models later.

Moreover, to better see Netflix's place in the Spanish market and the threat it poses to the current players, we must not forget the need to analyze the other forces that shape an industry, like the competition, the power of suppliers or the power of customers.

An analysis of Netflix's strengths, weaknesses, opportunities and threats in Spain shows a company with potential, but that faces a strong oligopoly, in an industry struggling with target audiences with rapidly changing consumption habits. Its weaknesses and threats run along the fact that its competitors of the Spanish markets have a tradition of more than three decades buying premium content, making contracts with the mayors and handling the rights for sports. Some of the Spanish competitors have the confidence of their subscribers because of their service and offer and, for now, have won the battles for rights and content.

Another barrier that any entrant in this industry must not forget about is infrastructure. Infrastructure works as another type of "raw material" essential for the activity of the company and it needs to be taken into account when thinking about suppliers. A good thing about OTT is that it works as long as there is an open Internet connection, and connectivity is nearly universal, getting better very fast, despite still lacking in some parts of the world.

However, it is still an issue to be considered. In Indonesia, for example, the country's state-run telecommunications provider, Telkom, blocked Netflix, citing permit issues [WOR 16]. Telecommunication companies have a distinctive advantage in that they already control the needed infrastructure and can mostly afford to take the risk of venturing into the content market. They own the "pipes" that run through the city into people's homes; any endeavor that gives them control over content that can be sent through the said pipes is an opportunity and a boost that keeps those pipes working.

The technological dimension is very important for an optimal user experience. This means that the OTT platform needs to take care of that aspect in the design of interface and the design of the user experience. It also means that the "pipes" must be good enough to support the image definition and provide a good service. It is important to avoid frustrations for the user in any point of the experience. Fluid reproduction, moving through content without problems, speed, high definition and good subtitles facilitate a smooth use of the technology.

Still an alien in the Spanish market, Netflix's strategy can be catalogued as strength and also as an opportunity when managing potential expectations and potential problems. For example, its interest in committing to national production in spite of still being free from the constraints of the European Directive 89/552 *Television without Frontiers* and its broadcasting European quotas has been seen as a step in the right direction toward the alliances with local industry. Any service entering a consolidated market will need to see cooperation as a strategy to start a good relationship with its suppliers. Meaningful relationships are essential for the creation of value and to be able to offer a quality product to engaged and potential consumers.

Hence, if we were to wonder if Netflix is a friend or foe for the Spanish competitors, after the analysis of competitors and the strategy of Netflix, we can conclude that it wants to be seen as a friend. It might help dealing with the piracy problem and it has signed alliances with telecom, service providers, producers and broadcasters.

However, Netflix could also be seen as a foe. The service seems to aim at an emerging, niche market formed by younger audiences open to online consumption, everywhere, anytime, mobile, comfortable and affordable, where they can be the center of the consumption experience. Such a market is an emerging one because of the blurred lines of its structure and the erratic behavior of its players, but it is attractive and expects growth. Legacy open television and pay television in Spain are still more consumed by a different – older for open TV and wealthier for pay-TV – target. For now, Netflix may not seem like a real foe in the oligopolistic market structure but it will probably be one against the smaller OTT services like *Filmin*, *WuakiTV*, *ADNStream*, *Filmotech* and similar. Like other new digital companies, Netflix offers three competitive advantages that have to do with customer intelligence, flexible pricing and Internet learning [SIM 16].

Change and innovation are constant in the media industry. As explained by Doyle [DOY 10, p. 432], "technological change is a constant feature in media industries and a number of studies have concerned themselves with how organizations have adapted their strategies to deal with convergence and growth of the Internet". Growth and change are signs of life and opportunities in the industry, and new highly efficient competitors will continue to emerge as long as there are opportunities.

The audiovisual industry works more as an amalgam of industries and new entrants might start to enter the market from unexpected sectors and industries. Entrants will not necessarily need to be better or offer bigger catalogues of content; they can start by offering services that are easier to use and less expensive than the existing options in the market. Incumbent competitors need to be flexible enough to handle the innovations that entrants bring into the market as an opportunity, identifying the strengths that could be transferred to their own existing model.

OTT video solutions cannot really be considered new anymore. More attention and money will be invested in these options and the market will continue to grow, intertwined with the more traditional media players and their own line of growth and innovation. As the market continues to mature, the pace of change will accelerate; will become cheaper and faster. OTT options had been looking for a share of the TV market in Spain for a long time, experimenting and trying to find the right business models and revenue models. Either a friend or foe to incumbent competitors, the arrival of Netflix, a strong brand, has served to bring the idea of legal SVOD to the forefront of people's minds.

2.4. Bibliography

[ALA 16] ALA-FOSSI M., LAX S., "The short future of public broadcasting: replacing digital terrestrial television with Internet protocol", *The International Communication Gazette*, vol. 78, no. 4, pp. 365–382, 2016.

[ANT 15] ANTHONY S., "Piracy laid the groundwork for Netflix's launch in Spain, says CEO", *Ars Technica*, available at: http://arstechnica.com/business/2015/06/piracy-laid-the-groundwork-for-netflixs-launch-in-spain-says-ceo/, June 2015.

[BAR 16] BARRET B., "Netflix's Grand, Daring, Maybe Crazy Plan To Conquer The World", *WIRED*, available at: http://www.wired.com/2016/03/netflixs-grand-maybe-crazy-plan-conquer-world, March 2016.

[COC 15] The Cocktail Analysis, "Televidente 2.0. Audiovisual Consumption trends in Spain (8ª wave)", available at: http://tcanalysis.com/blog/posts/televidente-2-0-audiovisual-consumption-trends-in-spain, February 2015.

[DOU 15] DOUGLAS P., *The Future of Television: Your Guide to Creating TV in the New World,* Michel Wiese Productions, Studio City, 2015.

[DOY 10] DOYLE G., "From television to multi-platform. Less from more or more for less?", *Convergence: The International Journal of Research into New Media Technologies*, vol. 16, no. 4, pp. 431–449, 2010.

[DOY 16] DOYLE G., "Digitization and changing windowing strategies in the television industry: negotiating new windows on the world", *Television & New Media*, vol. 17, no. 7, pp. 629–645, April 2016.

[FAH 16] FAHEY M., "Netflix vs Amazon: estimating the better deal", *CNBC*, available at: http://www.cnbc.com/2016/04/22/netflix-vs-amazon-estimating-the-better-deal.html, April 2016.

[FER 16] FERNANDEZ J., "Spain catches up with European VOD trend", *Rapid TV News*, available at: http://www.rapidtvnews.com/2016033042258/spain-catches-up-with-european-vod-trend.html#ixzz48BZ6BKjL, March 2016.

[GOD 15] GODLOVITCH I. *et al.*, Over the-Top players: market dynamics and policy challenges, European Union, available at: http://www.europarl.europa.eu/RegData/etudes/STUD/2015/569979/IPOL_STU%282015%29569979_EN.pdf, 2015.

[GON 15] GONZÁLEZ M., "Netflix en España, un mes después: ¿merece la pena renovar la suscripción?", *Xataka*, available at: http://www.xataka.com/streaming/netflix-en-espana-un-mes-despues-merece-la-pena-renovar-la-suscripcion, November 2015.

[HER 03] HERRERO M., *Programming and Direct Viewer Payment for Television. The case of Canal Plus Spain*, Media Markets Monographs, Pamplona, 2003.

[JEN 16] JENNER M., "Is this TVIV? On Netflix, TVIII and binge-watching", *New Media & Society*, vol. 18, no. 2, pp. 257–273, 2016.

[MED 15a] MEDINA M., HERRERO M., ETAYO C., "The impact of digitalization on the strategies of pay TV in Spain", *Revista Latina de Comunicación Social*, vol. 70, pp. 252–269, 2015.

[MED 15b] MEDINA M., HERRERO M., GUERRERO E., "Audience behaviour and multiplatform strategies: the path towards connected TV in Spain", *Austral Comunicación*, vol. 4, no. 1, pp. 157–176, 2015.

[NET 16a] Netflix Media Center, "Del Director Y Los Productores De Velvet Y Gran Hotel, Llega La Primera Serie Original De Netflix Filmada Y Producida En España", available at: https://media.netflix.com/es/press-releases/from-the-director-and-producers-of-velvet-and-gran-hotel-comes-the-first-netflix-original-series-shot-and-produced-in-spain, 31 March 2016.

[NET 16b] Netflix Investor Relations Website, available at: http://ir.netflix.com/long-term-view.cfm, 2016.

[OBS 15] Observatorio Jurídico de la Publicidad, "Sanción Por Incumplimiento De Los Límites De Tiempo Para Publicidad En Televisión", available at: http://observatoriojuridicodelapublicidad.com/sancion-por-incumplimiento-de-los-limites-de-tiempo-para-publicidad-en-television/, December 2015.

[POR 01] PORTER M., "Strategy and the Internet", *The Harvard Business Review*, vol. 79, no. 3, pp. 62–78, 2001.

[POR 08] PORTER M., *On Competition*, Harvard Business Review Press, Boston, 2008.

[ROL 15a] ROLFE P., "Spain Hits New High in Piracy, Study Says", *Hollywood Reporter*, available at: http://www.hollywoodreporter.com/news/spain-hits-new-high-piracy-780680, March 2015.

[ROL 15b] ROLFE P., "Netflix Launches in Spain with Star-Studded Event", *Hollywood Reporter*, available at: http://www.hollywoodreporter.com/news/netflix-launches-spain-star-studded-833232, October 2015.

[SIM 16] SIMON J.P., "The Digital Transformation New Players in the Media and Content Industry", *Telos*, vol. 103, pp. 1–17, available at: https://telos.fundaciontelefonica.com/DYC/TELOS/LTIMONMERO/DetalleArtculo_103TELOS_AUTINV2/seccion=1288&idioma=es_ES&id=2016031019050001&activo=6.do, 2016.

[SMA 12] SMALL O., "Reshaping the music distribution model: an iTunes opportunity", *Journal of Media Business Studies,* vol. 9, no. 4, pp. 41-68, 2012.

[VON 15] VONDERAU P., "The politics of content aggregation", *Television & New Media*, vol. 16, no. 8, pp. 717–733, 2015.

[WOR 16] WORO JUNIAR R., "Netflix blocked by Indonesia's top telecom provider", *The Wall Street Journal*, available at: http://www.wsj.com/articles/netflix-blocked-by-indonesias-top-telecom-provider-1453896220, January 2016.

3

The Resistance of Legacy TV Channels

Television has been, since the nineties, one of the most profitable media. The largest communication groups built their empires around this medium and, for years, it has provided their highest income. However, the economic crisis, the increasing supply and the proliferation of devices to access content have caused a shift in the nature of television and its business models.

In the last few years, public broadcasters, as well as private ones, have placed their content on the Internet. This has been a gradual process and not exempt from doubts and difficulties for the TV channels. Simultaneously, there has been a rapid increase in the use of the Internet. According to the last research done by EGM [EGM 16] (Estudio General de Medios), the Internet's penetration in Spain – the connection in the last few months – in 2016 was 77% while in 1996 it added up to only 1%. In 2015, the average daily use of the Internet was 102 min, which was surpassed by the TV with 238 min and the radio with 105 min.

This reality brings about important issues related to the television business model and distribution of content. The Internet can be a threat to television channels or it can become an opportunity for them [DEN 03]. According to Jin [JIN 13, p. 22], "for both network and cable television companies, finding a workable strategy in an increasingly digital

---

Chapter written by Mercedes MEDINA.

environment has been a significant issue, because they worry that the television´s role in the media landscape could diminish as Internet TV and social media, including YouTube, are taking a more important role as aggregators and distributors of information and entertainment content".

Dennis [DEN 06, p. 26] assures that "television has begun to redefine itself and its role in the media family. It is an industry, a technological platform, a content creator and distributor of information and entertainments and advertising". Some authors suggest that the term "television" needs to be revised [RAN 03] [TOL 09], because "the way television is produced and delivered, as well as how and where it is watched, is changing" [ROS 04].

However, some authors express their doubts about the capability of new media to develop their own business independently from traditional television. According to Murphy [MUR 11, p. 59], "television is central to new media". Harries [HAR 02, p. 207] questions "if new media can find its own voice, its own vision, detached from the traditions of the cinematic and the televisual?"

Questions remain unanswered regarding which business model will prevail, where the money will come from (from advertisers or paying users), and what the most used device will be.

Amid this situation of uncertainty, in the following pages, we want to offer some ideas that are based on the data about what the private TV broadcasters financed by advertising have given to the audiovisual market in Spain. In other words, what would have happened if private television companies did not exist and if there was no free private television in the future? The reflection stems from the risks that said disappearance could bring and we will look at the opportunities that its existence offers to the audiovisual market, economy and society.

3.1. The engine of audiovisual market

The television market depends on two sectors, the advertising sector and audiovisual production. These sectors would not be able to develop without the existence of commercial television.

Since the beginning of private television in Spain, the main income of the television sector has come from advertising, especially if we consider that until 2010 public television had advertising too (see Figure 3.1). The loss of advertising in the national public broadcaster helped increase revenues of private television, but the influence of the economic crisis affected them and the decline in these revenues continued. As a result, advertising revenues in 2011 were lower than those from 2000 and only in 2014 was there a small upturn.

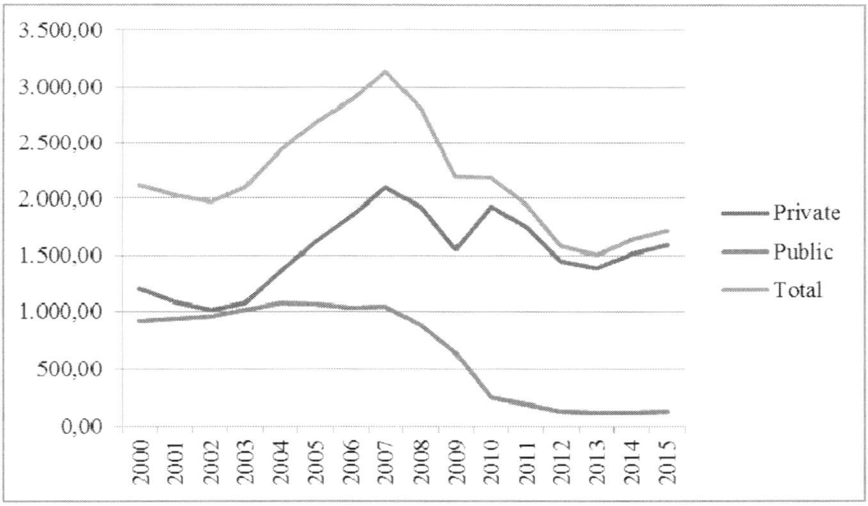

Figure 3.1. *Evolution of TV advertising revenue (million €)*[1] *(source: CNMC; last update 27/10/2015, data extraction 24/05/2016). For a color version of this figure, see www.iste.co.uk/medina/audiovisual.zip*

As shown in Table 3.1, more than 80% of advertising revenues are in the hands of two audiovisual groups, Mediaset and Atresmedia. Their most important source of revenue is advertising from their television channels, which accounts for 95% of sales. Therefore, their core business is television distributed through the terrestrial digital system.

1 Advertising revenues refer to net income, after deduction of the discounts and commissions charged by companies selling the advertising.

	2012	2013	2014
Mediaset España	685.91	637.28	707.04
Atresmedia	575.57	653.98	705.52
Total private TV	1,261.48	1,291.26	1,412.56
Total advertising revenue	1,576.74	1,503.73	1,634.86

Table 3.1. *Advertising revenues of private audiovisual groups (millions €) (source: CNMC; last update 27/10/2015, data extraction 24/05/2016)*

Unlike revenues from advertising, those of pay television are less sensitive to the economic situation. Their evolution is more stable. If we compare the evolution of these two types of revenues, it is possible to discover three historical moments where payment reaches and exceeds advertising revenues: in 1999, in 2004 and in recent years, from 2012 to 2014 (see Figure 3.2).

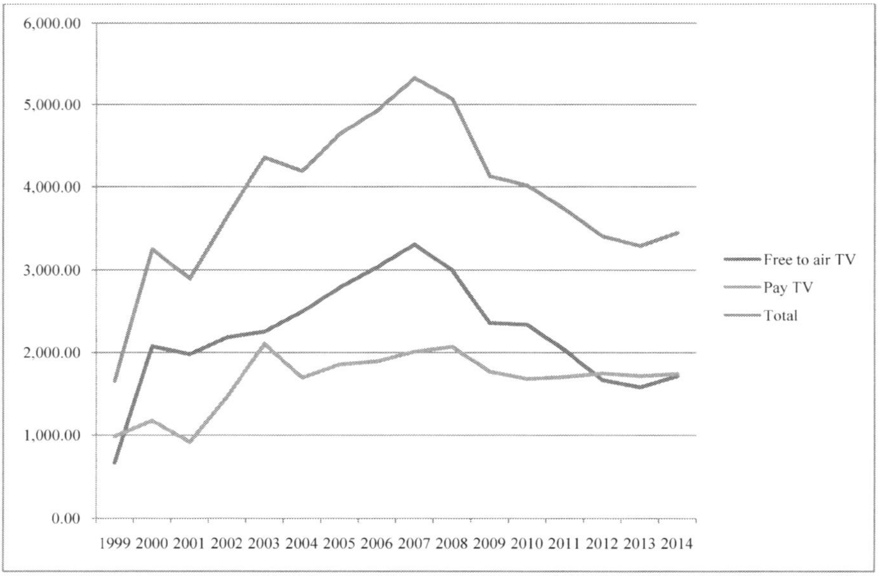

Figure 3.2. *Evolution of TV revenue and pay-TV (million €) (source: CNMC; last update 27/10/2015, data extraction 24/05/2016). For a color version of this figure, see www.iste.co.uk/medina/audiovisual.zip*

During the last period of economic recession, revenues from pay television exceeded those of free TV financed by advertising for the first time. In 2012, free TV billed 1,665.8 million euros (17.9% less than in 2011) and pay-TV services totaled 1,746.8 million euros, an increase of 2.5%. These data lead to question the profitability and permanence of advertising revenue as sustenance of the television industry.

However, large advertisers still consider television as the cheapest and most effective way to reach the public. This explains that although the Internet is the fastest growing medium in advertising investment, television continues to attract greater share of advertising investment, as shown in Figure 3.3. From 1999 to 2013, investment in Internet has increased by 20%, but it has not been at the expense of television or radio or outdoor advertising, but media like newspapers, magazines and movies.

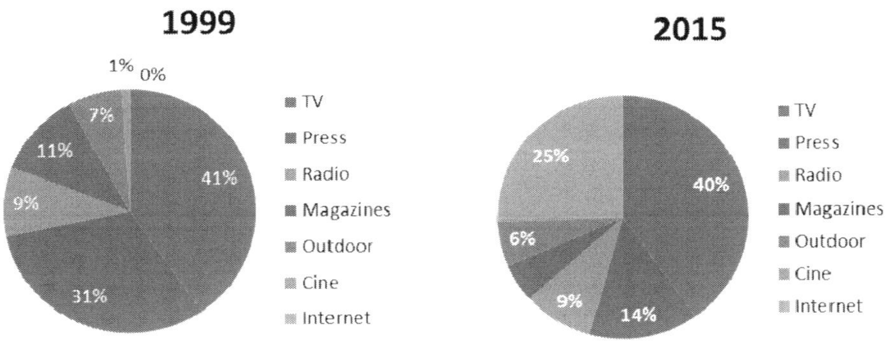

Figure 3.3. *Advertising investment (real estimated) (million €) (source: [INF 15]). For a color version of this figure, see www.iste.co.uk/medina/audiovisual.zip*

Over 25 years of existence, private TV networks have not only provided scope and coverage to the advertising sector, but also enabled the development and maturity of the Spanish audiovisual production sector. Besides being generators of content, they have motivated the formation of large production companies with international prestige [DIE 08].

In Figure 3.4, we can see the evolution of a number of independent producers in Spain in recent years.

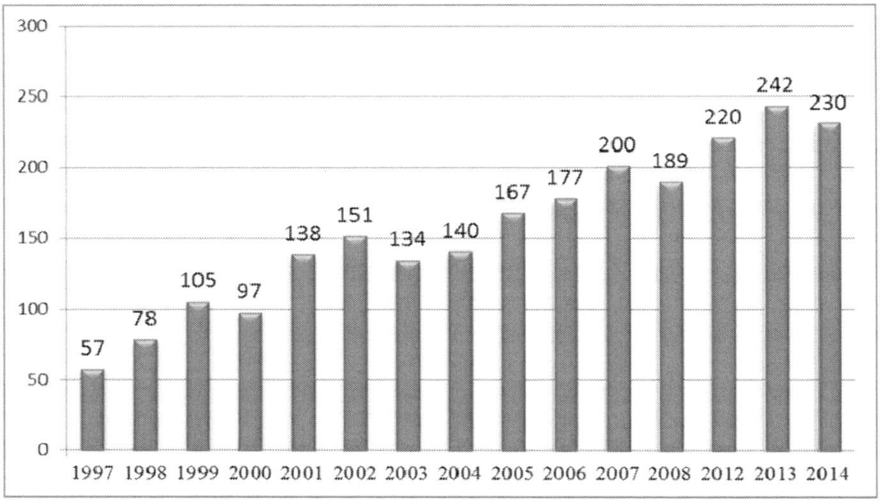

Figure 3.4. *Number of independent producers (source: FAPAE)*

Even though many of these producers are small companies with little activity, private television has managed to generate a business landscape that has expanded the number of jobs in the audiovisual sector and has allowed the development of creative projects that have been exported to other countries.

It is possible to see to what extent broadcasters support and need independent production by extricating the list of production companies who have been hired by TV channels. It also helps us to see which of those hundreds of companies have more structure and stability, thanks to their work with the networks. The following table highlights the 25 most watched programs in Tele 5 and Antena 3 TV in 2015 that were produced by independent producers.

Program	Producer	%	Rkg	Program	Producer	%
Eurobasket		14.5	**1**	Futbol:champions league		14.8
Fútbol: copa del rey		14.2	**2**	El peliculón		13.6
Baloncesto:eurobasket		13.8	**3**	Prorroga futbol		12.7
Motociclismo		13.3	**4**	Alli abajo	Plano a plano	12.0
Futbol:supercopa		12.9	**5**	Ell debate decisivo		11.3
Cine		12.0	**6**	Velvet	Bambú	10.3
El principe	Plano a plano	11.5	**7**	Vis a vis	Globomedia	9.7
La voz kids:final	Boomerang	11.1	**8**	Bajo sospecha	Bambú	9.5
GH VIP:express	Zeppelin	11.0	**9**	El hormiguero 3.0	7 y acción	9.4
La voz kids	Boomerang	10.5	**10**	Mar de plástico	Boomerang	9.1
La voz	Boomerang	10.4	**11**	Supercopa europa		8.8
Futbol:amistoso		10.2	**12**	Supercopa europa		8.4
GH VIP:express	Zeppelin	10.1	**13**	En tierra hostil		8.1
La voz kids	Boomerang	10.0	**14**	Especial velvet	Bambú	7.5
La voz kids:batallas	Boomerang	9.7	**15**	Sin identidad	Diagonal	7.5
La voz kids expres	Boomerang	9.7	**16**	Tu cara me suena		7.5
Informativos T5 21h.		9.6	**17**	Algo que celebrar	Secuoya	7.4
Anclados	Globomedia	9.4	**18**	Formula 1		6.8
La que se avecina	Contubernio	9.4	**19**	Casados a primera vista	Boomerang	6.7
Los nuestros	Multipark Ficción	8.9	**20**	Me resbala:especial terror		6.7
La voz expres	Boomerang	8.7	**21**	Futbol:copa del rey		6.6
Post motociclismo		8.5	**22**	The flash		6.6
La voz:directos	Boomerang	8.5	**23**	Multicine 2		6.5
Supervivientes		8.4	**24**	Antena 3 noticias		6.5
Aqui paz y despues gloria	Mandarina	8.3	**25**	Cine		6.5

Table 3.2. *Ranking of most watched shows Tele 5 and Antena 3 (maximum audience) 2015 (source: [KAN 15])*

Among the most watched programs in Tele 5, 13 are produced by independent producers. Specifically, Plano a Plano, Boomerang, Zeppelin, Contubernio, Multipark and Mandarina have produced for private channels. Tele 5 usually has a programming strategy focused strongly on the success of a few programs. Overall, its most watched programs belong to the genre of reality television. In recent years, they have had *Big Brother* and *The Voice*. It has also spent years making a commitment to sports other than soccer, such as motorcycling and basketball.

In contrast, of the 25 programs of Antena 3 TV, only eight came from independent producers. In addition to Boomerang, Globomedia and Plano a Plano, the network has also hired programs produced by Bamboo, Sequoya, 7 y acción and Diagonal. The programming strategy at Antena 3 favors more varied programs for varied audiences. Its bet on airing soccer has been more intense than Tele 5's and it counts on reaching large audiences through series, comedy shows and movies.

Some of these companies produce more than 10 programs a year and their programs have remained on air for years and have even been sold to other countries. For example, according to FAPAE Annual Report [FAP 14], *El Príncipe* [The Prince] has been sold in most of Latin America and other countries like Croatia, Italy and Portugal. *El tiempo entre costuras* [The time in between] by Boomerang TV, for Antena 3, which was followed in Spain by more than four million viewers, was sold in the Latin American market and in Asian countries like the Philippines. Series produced by Bamboo Productions such as *Velvet*, followed by 4.8 million viewers, and *Gran Reserva* on TVE found good reception in some European countries and Mexico.

Other production companies are younger and produce fewer programs for national channels, which results in their turnover and business structure becoming more limited.

3.2. Economic potential

Employment rate and economic results are indicators of the good performance of an industrial sector. Below, we will see the volume of these indicators powered by the private television companies in Spain.

Another indicator of the contribution of one sector to the national economy is the number of employees. Table 3.3 shows the number of employees of the audiovisual market in the 2000s. It is possible to observe how from 2004 to 2008 more than a thousand jobs were lost; no doubt as a result of the crisis. Many companies, especially public ones, had to reduce their workforce and some of them even closed.

2004	23,081
2005	23,951
2006	23,079
2007	24,651
2008	21,979

Table 3.3. *Number of employees in the audiovisual sector (source: CNMC)*

Nevertheless, after the worst years of economic crisis, the two private audiovisual groups generated wealth, expressed in number of employees and benefits. As for the number of employees, although they may not have reached the levels of the public sector, together they can sum up to more than 3,000 employees. In 2013, the Mediaset Group Spain had a total of 1,308, in 2014, 1,274 and in 2015, 1,266 employees, located mainly in Spain, and Atresmedia in 2013 had 1,829, in 2014, 1,806, and in 2015, 1,917.

As to the economic and financial figures, the two groups have generated revenues and profits in recent years. They managed to overcome the consequences of the economic crisis and emerge as established and creditworthy companies. Table 3.4 summarizes the main economic data for these companies in 2013 and 2014.

		Mediaset España	Atresmedia
Operating income	2014	932	883
	2015	972	970
Benefit	2014	55	47
	2015	166	99

Table 3.4. *Economic and financial results (millions €) (source: CNMC; last update 28/10/2015, data extraction 31/05/2016)*

The turnover of the two private audiovisual groups in these years exposed a sum of more than € 2.5 billion. However, from the point of view of profitability, Mediaset is more efficient at generating wealth than Atresmedia, which has a higher proportion of costs and greater debt.

We cannot ignore the fact that these groups have been benefited by Law 10/2005 June 15th, on urgent measures to boost DTT, Liberalisation of Cable Television and Promotion of Pluralism, which allowed the absorption of the TV channels La Sexta, from Imagina-MediaPro Group by Atresmedia and Sogecable's channels by Mediaset Spain. It seems like a paradox that a law that promotes "pluralism" favors market concentration.

Television is the core business of these groups. They have created a consortium of partner companies around this medium. Mediaset's strategy has focused on vertical integration and, within the Spanish subsidiary, there are companies like Publiespaña, Publimedia management, Grupo Editorial, Tele 5 Cinema, Conecta 5, Premiere Megaplex and Integración Transmedia. In the case of Atresmedia, growth operations have been aimed at creating a multimedia group that includes television, radio, music companies, film and audiovisual production.

Although both have invested in online business and have diversified into new screens and the distribution of their content via the Internet, their main source of income remains the traditional television. In 2013, Atresmedia made 9 million euros with online advertising, which represented 1.7% of the network's advertising revenue. These figures show that these services are still experimental and income is residual.

The exploitation of the new online advertising market will come when Kantar Media makes available market data as complete as those related to the traditional television consumption. Atresmedia wants to be prepared for the technological transformation and has created a department, Atresdata, to exploit and market the data they have coming from mobile devices and online access to their content.

3.3. Contribution to society

Like all media, television finds its purpose insofar as it is accepted by its audience. Therefore, figures regarding the time spent watching programs and other audience data are fundamental measures to assess their contribution to society. But as companies that produce intangible assets, the quality of their programs and their social responsibility are also indicators of the ethical and moral contribution that television makes to society. Consequently, in this section we will focus on the consumption times of television and other actions that private televisions in Spain have carried out as part of its Corporate Social Responsibility.

Despite the increased consumption of online content, hours of television consumption have remained at 4 h over the last 5 years, as shown in Figure 3.5.

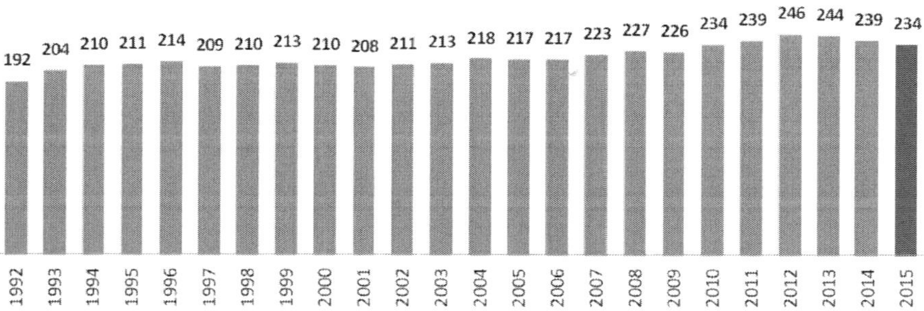

Figure 3.5. *Annual evolution linear TV consumption (min. ind/day) (source: Kantar Media)*

The population groups that watch less television and whose downward trend continues over time are children and teenagers. While other population groups have increased their consumption time since 2001, adolescents have remained at about 2 h and 10 min with little variation, as shown in Figure 3.6.

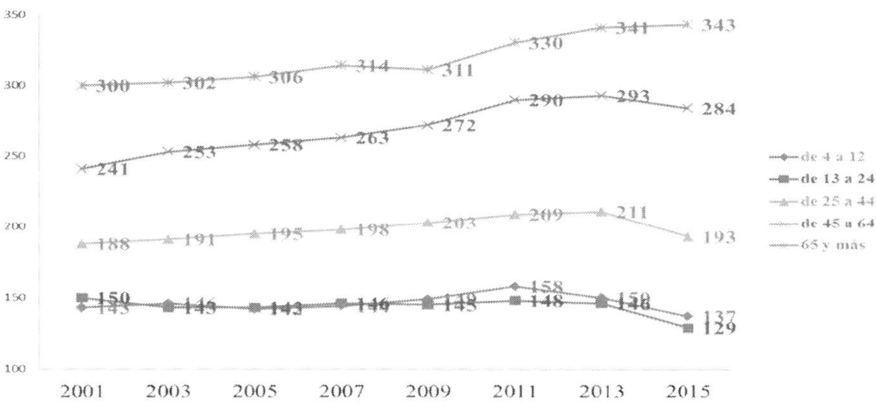

Figure 3.6. *Evolution of TV consumption according to age segments (average daily minutes per person) (source: TNS and Kantar Media). For a color version of this figure, see www.iste.co.uk/medina/audiovisual.zip*

The biggest advantage of the digital terrestrial television is its capacity to reach large audiences. Despite the current fragmentation of the market, for large advertisers and opinion makers it remains an attractive medium due to its effectiveness at attracting many viewers.

Moreover, audience measurement based on audiometry has a backup of years of experience and prestige. Despite its limitations [MED 03], the samples are representative of the population, have a body of auditors and work as a guaranteed currency for all agents involved in the sector. It also provides the value of anonymity to panelists, an issue of particular importance in the current measurement of online devices.

Among the most watched programs in recent years, we have already pointed out that soccer matches have a special place, matches whose rights are very expensive and for which the direct competitors are pay-TV channels like *Movistar +*, cable operators and channels such as *BeINSport* and *YouTube*. There are other genres such as news, national series, reality and entertainment shows that attract less competition. The content that turns television into an essential staple in home audiovisual consumption will be that which is aired live, the national series productions and the news when sources and information are verified and there is an international coverage. Other genres such as American series or motion pictures will probably remain the favorite offerings from online audiovisual services like *ADNStream*, *Wuaki* and Netflix.

Another important aspect of private television companies is the Corporate Social Responsibility activity complementary to its main social function of informing and entertaining.

For years, both Tele 5 and Antena 3 have run a social campaign of deep impact on society. “Twelve months, twelve causes” began in 2001 and is now called “12 months”. With this campaign, Tele 5 endorsed a social cause year after year and month after month. The last campaign of 2015 was headed to the protection and defense of women and the fight against child poverty. Under the name of *Doy la cara* [Stand up], famous faces from the world of arts and literature lent their voices to these causes.

Following Tele 5, in 2008 Antena 3 launched the campaign *Ponle freno* [Stop]. It was the same person who launched the campaign in Tele 5, Javier Bardají, who influenced Antena 3 years later. This campaign was intended to denounce, with the cooperation of citizens, the black spots on Spanish roads where most car accidents occurred. The money raised in a race organized in 2015, where 20,000 runners participated, went to the victims of car accidents.

In addition to carrying out these campaigns with high social impact through television coverage and with the brands of these TV channels, the groups donated part of their income to charitable causes. Specifically, Antena 3 allocated more than 1 million euros from 2013 to 2015 to NGOs as "Acción contra el Hambre" (Action against Hunger), the Food Bank, Greenpeace, the Foundation for Assistance to Drug Addiction or "Tolerancia Cero" (Zero Tolerance) against gender violence. According to the annual report of Tele 5, the company gave more than 48 million euros to various NGOs in the same period. This amount includes the advertising spaces assigned to social causes.

Finally, it is worth noting, as an intangible value of these companies, the awards they have received for the quality of their programs or for the careers of their professionals. In 2015, Mediaset received the Vitoria FesTVal award for its show "GH VIP" as the best entertainment program and the "Antena de Oro" Award for its series "Lo que se avecina". Stars linked to Tele 5, such as Maria Teresa Campos, Pedro Piqueras or Iker Jiménez also received several awards.

The Atresmedia group also received some awards in 2015 for the work of its presenters such as Matias Prats, Wyoming and Jordi Evole; the "Ondas" award for programs like "Tu cara me suena" and "Vis a vis" and the FesTVal Special award for the special program on the elections and their comedy "Allí abajo".

Therefore, we can conclude that these private media groups, thanks to their free TV channels, have engendered an audiovisual industry that benefits both advertising agents and audiovisual producers. They have earned the favor of viewers through its entertainment and informative programs. They have generated jobs directly in the industry and have contributed to many others, whose work has been recognized by other social actors and their presenters have served for important charitable causes. Undoubtedly, in 20 years of existence, they have also shown weaknesses, mistakes and complicated economic situations. But after this exhibition, it is time to think about the nature of television and its contribution to society, industry and professionals.

3.4. Bibliography

[CNM 15] CNMC [National Commission of Market and Competition], CNMC Data, available at: http://data.cnmc.es/datagraph/jsp/inf_anual.jsp, 2015.

[DEN 03] DENNIS E., "Prospects for a big idea – is there a future for convergence?", *International Journal of Media Management*, vol. 5, no. 1, pp. 7–11, 2003.

[DEN 06] DENNIS E., "Television's convergence conundrum: finding the right digital strategy", *Television Quarterly*, vol. 37, no. 1, pp. 22–26, 2006.

[DIE 08] DIEGO P., PARDO A., "La ficción televisiva en España: reinvención del género y adaptación de la industria", in MEDINA M., FAUSTINO P. (eds), *The Changing Media Business Environment*, Formalpress, Lisbon, pp. 43–58, 2008.

[EGM 16] EGM, Marco General de los Medios en España, AIMC, Madrid, available at: http://www.aimc.es/-Datos-EGM-Resumen-General-.html, 2016.

[FAP 15] FAPAE (Federación de Asociaciones de Productoras Audiovisuales de España), *Memoria annual*, available at: http://fapae.es/informes/memoria_anual, 2015.

[HAR 02] HARRIES D., *The New Media Book*, BFI, London, 2002.

[INF 15] INFOADEX, Estudios de la inversión publicitaria en España, available at: http://www.infoadex.es/estudios.html, 2015.

[KAN 15] KANTAR MEDIA, Anuario Audiencias, available at: http://anuario.kantarmedia.es/, 2015.

[MED 03] MEDINA M., "Audiencias y programas de televisión. Alternativas a la audimetría", in ARRESE A. (ed.), *Empresa Informativa y Mercados de la Información*, EUNSA, Pamplona, pp. 227–258, 2003.

[RAN 03] RANGONE A., TURCONI A., "The television (r)evolution within the multimedia convergence: a strategic reference framework", *Management Decision*, vol. 41, no. 1, pp. 48–71, 2003.

[ROS 04] ROSCOE J., "Multi-platform event television: reconceptualizing our relationship with television", *The Communication Review*, vol. 7, pp. 363–369, 2004.

[TOL 09] TOLETTI G., TURBA L., "Sofa-TV: the new digital landscape", *International Journal of Digital Multimedia Broadcasting*, vol. 2009, pp. 1–8, 2009.

4

The Public Service Heritage

Regular television broadcasting services began in Spain on 28 October 1956. As in other European countries, the television broadcasting came under the ownership and management of the Government. Unlike other European public television corporations, the Spanish public broadcaster benefited from dual funding, being financed by both state subsidies and advertising.

Commercial television arrived in Spain in 1990 much later than in other European countries, following the publication of the Private Television Act (10/1988) which regulated private television companies. Under this Act, three television licenses were initially granted, two of them to free-to-air channels Telecinco and Antena 3 and one to the pay-per-view channel Canal +.

The introduction of new channels to the audiovisual market caused, among other things, a decline in audience ratings and an increase in costs for the public broadcaster and within 15 years, RTVE had accumulated a debt of 7,850 million euros. Its economic and financial situation became untenable, while its public service identity and legitimacy were questioned [RIC 94].

However, this crisis has not only taken place in Spain, but also in other European countries, and in the United States as well. According to Campos-Freire [CAM 13], this stagnation manifests itself in financial difficulties, lack of flexibility in its management and adaptation to the environment, since in most cases, the structures are too large and rigid and many are

Chapter written by Mercedes MEDINA.

facing serious inefficiencies while trying to adapt to the new digital ecosystem.

The need to transform the public operator was pressing, but the reforms carried out by various governments up to that point had been few and inconsistent.

At present, the RTVE Group is marked by a complex legal structure made up of RTVE Corporation and two public limited companies (TVE and RNE) and is geographically dispersed, with 17 TVE regional production centers and 62 RNE regional, provincial and local radio stations, in addition to 18 foreign correspondent offices. In terms of its business units, the Group is comprised of three general-interest and five thematic television channels, six radio channels, one record label, the RTVE Orchestra & Choir, and the RTVE Institute for Audiovisual Training. It also broadcasts the pay-TV-channel *Docu TV* as well as the international channels TVE Europa (for Europe and Asia) and TVE America via satellite. The latter two channels are part of *TVE Internacional*. In addition, RNE runs four nationwide radio stations – RNE1, Radio Clásica, Radio 3 and *RNE 5 TodoNoticias* – and one for Catalonia, RNE 4.

After a lengthy consultation process with experts and influenced by the British and French models, the Spanish government undertook a new reform of RTVE in 2006. Thus, the Law of State Radio and Television (17/2006) was published, the main objectives of which were to achieve political independence and efficient economic management. Amongst the changes introduced to achieve these goals, the most effective ones have been changes in the governance of the public corporation and the removal of advertising as a source of revenue from 2010 onwards.

A survey run in 2009 concerning the trends in the audiovisual sector in Spain, carried out by the "Public Sector-Private Sector" research group from the business school of the University of Navarra [IES 09], revealed various views about the future of TVE. The survey included participants such as public and private television managers, telecommunication managers, technological manufacturers, advertising agencies, public regulators and audiovisual producers and experts.

More than 80% of interviewees felt that new audiovisual regulation was needed. However, there was no unanimous view on the abolition of advertising on TVE: 53.9% (mainly the public operators and media agencies) viewed the measure as inadequate; however, 71.4% supported the decision. The latter belongs to the private operators and technical manufacturers. According to most of those interviewed, this measure would lead TVE to offer different programs than its competitors and it would become more independent of audience ratings. Furthermore, half of them thought that TVE would downsize and externalize most of its productions.

In this chapter, we will focus on the origin of the crisis of public television in Spain, the current challenges faced by the corporation and the proposals to ensures it survives in the future.

4.1. The crisis of the Spanish public television

The current crisis at RTVE is both economic and a crisis of political independence. The decline in advertising revenue and the lack of availability of state aid, coupled with an increase in production costs and chronic overstaffing since 1990, has resulted in increased debt for the corporation. Furthermore, whilst TVE's identity as a public service broadcaster has been undermined since the creation of commercial television companies, the lack of independence from political power truly triggered the crisis.

As in the rest of the European countries, from 1956 onwards, RTVE's economic situation had been regarded as "miraculous" due to the lack of financial and material resources [PER 65]. Consequently, the arrival of advertising in 1958 put an end to the problem [PAL 05]. Advertising revenue grew exponentially from 1959, growing from 4 to 16 million pesetas in 4 years, 524 million in 1963, 4 billion in 1969 and 8 billion pesetas in 1975 [BUS 07]. Revenue was also obtained with the introduction of the television license fee for television owners, which ranged between 300 and 500 pesetas per year, a high amount for the salaries of the time. However, the license fee was definitively abolished in 1965 through the Act 103/1965 of 21 December 1965.

In 1980, almost 30 years after the birth of State television, the Radio and Television Statute (Act 4/1980) was published, which provided four types of funding through (1) subsidies allocated from the National Budget, (2) marketing and sale of its products, (3) limited participation in the advertising market, and (4) a television license fee payable by owners of color TV sets only. In those years, RTVE was self-funding for the first time, with enough advertising revenue to fund it and little need to resort to state aid.

The year 1983 saw the arrival of regional TV channels in Spain. While these too were public channels, they also competed for advertising with the only channel then in existence. Despite this competition, the advertising revenue of RTVE did not decrease [MED 10]. However, in 1989, the first set of commercial television channels was created. As a result, RTVE experienced a decrease in advertising revenue and an increase in debt, which continued to grow in subsequent years.

By the early nineties, spending started to exceed revenue. This was in no small way due to the maintenance of excessively bureaucratic and inefficient decision-making systems as well as a complex structure with high fixed overhead expenses. One of the biggest problems is the large number of staff. In 2003, Antena 3 had 1,600 employees, Tele 5 had 730, and TVE had 6,029. TVE was clearly overstaffed. Along with the overstaffing issue, another concurrent problem was employee absenteeism. According to the Spanish Court of auditors report [SPA 06], the absenteeism rate among management staff reached 14.45% in 2004, while general service workers had an absenteeism rate of 13.97%. One of the most significant characteristics of the bureaucratic apparatus of TVE was the sheer number of management staff. Among 6,122 employees, 4,170 were managers and technicians and 952 were personnel of administration, operations and sales.

For more than 30 years, TVE enjoyed a dominant market position that allowed the broadcaster to grow and diversify without worrying too much about its profit and loss account. However, the appearance of commercial television channels shifted TVE from the leading position and disturbed the balance of its accounts.

Private television channels, which fulfilled a similar entertainment function and offered similar products, attracted viewers and drove the cost of programs up. These problems were also compounded by the emergence of strong fully-fledged production companies and an increase in the cost of programs. The cost of television would rise again later on due to the need to invest in the transition to digital television and the development of new digital media.

As mentioned in the Court of Auditors report [SPA 06, p. 85], "the historical problem of an inefficient use of staff, which forces TVE to excessively outsource programs, thus originating a double cost for the public broadcaster as a result of maintaining its own staff inactive, that is to say, without tasks to be developed".

During the 2001–2004 period, in-house programs remained unchanged in relative terms, whilst the percentage of resources allocated to independent production increased, from 15% in 2001 to 26% in 2004, representing an increase of 20 million euros, to the detriment mainly of mixed productions, in which, TVE participates along with other third parties external to the Group, which underwent a reduction in resources and, accordingly, dropped from 33 to 23% of total costs. TVE central services in Madrid outsourced programs to 44 production companies in 2002, 40 in 2003, and 61 in 2004.

Increasing costs and a full in revenue generated a debt from 1990 onwards, which increased year-on-year until it reached 7.85 billion euros. The following graph shows the TVE's deficit and growing debt, which placed it in a situation that warranted an urgent response.

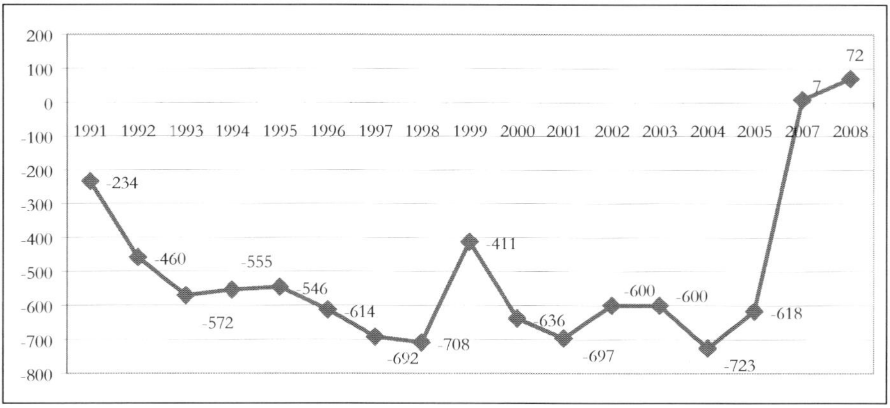

Figure 4.1. *RTVE results (1991–2005) in millions of euros (source: Bustamante [BUS 07, p. 229])*

Despite the implementation of reforms, the debt did not decrease significantly until 2007. Small inroads were made in 1999 as a result of the sale of Via Digital (a digital pay-per-view TV channel), the segregation of Retevision (the airwave distributor) and a reduction in VAT.

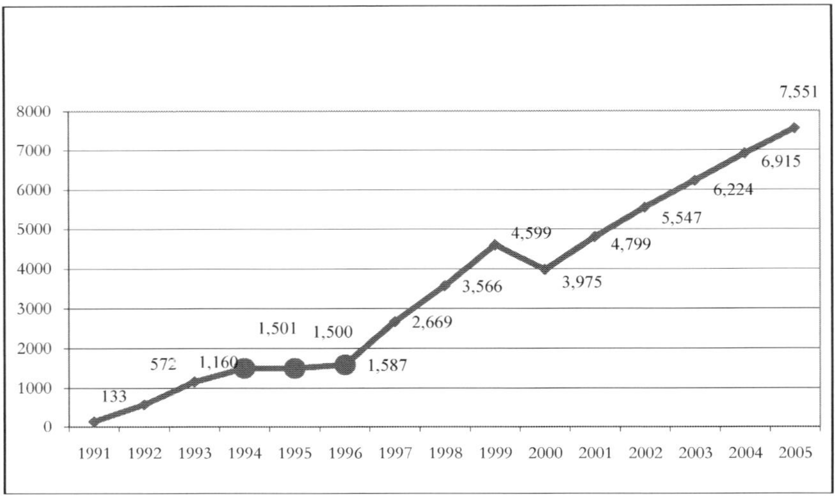

Figure 4.2. *RTVE annual debt (1991–2005) in millions of euros (source: Bustamante [BUS 07, p. 229])*

The other big problem for RTVE is political independence. The management of TVE was dependent on the public administration and conditioned by the political party in power at the time [ALM 10]. No government had ever attempted to get rid of such an influential medium. Additionally, any solution to the economic deficit would necessarily have involved staff reduction and risked a confrontation with one of the most critical sectors of society, the trade unions, which no government wished to offend [BUS 89].

The political crisis indicated above stemmed from the close relationship RTVE always had with the nation's Government. The short term appointment of directors-general and presidents of the corporation exacerbated the difficulty in developing a strategic plan for the broadcaster. Most of the directors-general remained in the post for less than 3 years, leading to a degree of instability in the directorship of RTVE and making it difficult to implement long-term strategic plans for its future.

The desire to pander to public interests prevented a solution to deal with RTVE's economic deficit being found. Fear of negative public reaction to spending cuts led to indecisiveness and explains why some governments maintained economically unviable structures with programs that did not meet audience demands, making it impossible for the organization to achieve the institutional principles it is based on. Solving the funding problems of public television services required consensus among the political parties and the undertaking of measures which, albeit painful in the short term, would guarantee the economic viability of public service television channels in the long term.

4.2. The current challenges

Reform of the financing of RTVE arrived with the Law 8/2009 passed under the socialist government. With this law, RTVE was renamed a Corporation and the election of the President and the Board of Directors was reformed in order to keep the company independent of political influence. As concrete measures, advertising was removed and it was stipulated that private television stations would pay 3% of their income to cover this gap in financing, pay-TV stations 1.5% and telecom operators 0.9%, in addition to the fees already paid by radio and television stations for the use of the radio spectrum. Telcos protested on several occasions against having to fund the Public Corporation, but the European Commission rejected their complaint [EFE 13].

To achieve efficient management, they had to cut jobs. There were 9,317 employees in 2004, but this was reduced to 6,517 in 2014. Also, an armored budget of 1,200 million euros was passed to spend between 2012 and 2014 and a limit was set to spend 10% of the budget to acquire sports rights.

To attain editorial independence, the Law 17/2006 of June 5th on State-owned Radio and Television established that the Congress had to appoint the President for their merits and for 6 years. The Board of Directors would consist of twelve members, eight elected by the Congress and four by the Senate. These measures were again modified by Royal Decree 15/2012, which stated that nine councillors had to be elected by two-thirds of both chambers (Congress and Senate) in the first round, and if this was not possible, in the second round, they would be elected by an absolute majority.

In October 2012, the European Broadcasting Union (EBU) coordinated a study on the challenges that public television in Europe faces. The main objective was to identify the elements that make the public service media indispensable for audiences of the 21st Century. After a year of study and research, in late 2013, the report EBU *Vision 2020*[2013] was published. It states 10 recommendations for public service media to maintain their legitimacy in the audiovisual sector. Among other issues, it recommended (1) a better understanding of the audience and an increase of the engagement, (2) to provide quality and diverse content (3) to be a source of relevant and reliable information, (4) to be a benchmark for younger audiences, (5) to develop an institution that is open, collaborative and directed toward innovation and (6) to justify its nature as a public service medium.

To achieve a better relationship with its audience, RTVE opened a Facebook account in 2009 and used it to receive suggestions from its users. Moreover, through its website, it is possible to access all its programs, both current and archived, and to view the popularity of downloaded videos. In 2007, RTVE created a figure called “defensor(a) del espectador” [defender of the viewer], which collects the complaints and suggestions of viewers in the monthly program “RTVE responde” [RTVE answers].

In an effort to eliminate barriers for people with hearing disabilities, RTVE developed a plan of subtitling and it became a priority for the corporation. In 2012, TVE1, the first channel of the network, had captioned 75% of its programming and TVE, the second channel, had captioned 77% [RTV 15].

Moreover, the rights of different social, political and labor groups to access the content of the corporation is regulated to ensure pluralism and public service. However, some authors conclude that the presence of different social groups in the newscast is still weak [CAL 08].

Still, it is not uncommon to find news that put into question the commitment to pluralism and independence. For example, changes in the directive team of the news section sparked protests by workers of the network, who demanded real independence in their work [ELD 14, FER 14].

As an effort to serve disadvantaged communities, we can mention that RTVE wanted to give special coverage and reception in their programs to the immigrant population. It developed programs for immigrants such as

Destination Spain and *Babel*, where immigrants are the main characters [HER 11]. RTVE has also developed programs committed to social groups like *Capacitados* [Empowered] in TVE 2, a program that encourages people to get in someone else's shoes and live like those affected by a physical disability. Their sports channel also broadcast the para-Olympics.

TVE has shown that it is possible to do quality popular programs with values, without losing the appeal for the audience. Examples of this are the series *The Mysteries of Laura* from the Boomerang production company, which was sold to the US (NBC) and other countries such as Argentina, Russia, Holland, Italy and Turkey, and *Cuéntame* [Tell Me], sold to Mexico, Portugal, Italy, Argentina and Puerto Rico. Moreover, in addition to its presence in social networks with the suggestion box to listen the audience' suggestions, and the success of participatory formats such as "Master Chef" and the children's version "MasterChef Junior", we can note the attention the public company has given to young entrepreneurs. Through the program *CódigoEmprende* [Start-up Code], contestants can turn a project into a company with the advice of experts, if they convince the jury and win the prize of 100,000 euros.

TVE also encourages domestic production. In 2009, the corporation spent more than a quarter of its output on works by independent producers, for example, the *Cuéntame* series by the group Ganga Productions, *Isabel* by Diagonal or *Gran Reserva* by Bamboo. Moreover, in 2013, it broadcast a total of 283 Spanish films, which represented 92.5% of the Spanish film production broadcast on television.

In order to become an information reference channel, RTVE has developed a 24/7 news service through the Web and mobile app "24 horas" [24 hours]. Thanks to the network of 16 correspondents in America, Europe, Morocco, Jerusalem and China, its international information is almost always first hand.

Its innovative approach can also be seen on its website and online services [MAN 11]. RTVE allows access to all content and live feeds via any device and developed "the red button" to access all interactive content and services from smart TVs. Among the available options, it is possible to watch programs, share content via email, *Twitter* or *Menéame*. Through the apps of their programs, like "MasterChef", "Cuéntame" or "Isabel", the user can interact with the programs or capture 30 seconds of the live program and share it on social networks. In 2013, it developed a Media Lab to experiment with new transmedia products and explore new narratives. The Lab created games and applications for mobile devices about the series "*Aguila Roja*" and a map of characters and a monograph of the battles took place up to the conquest of Granada for the series "*Isabel*"[COS 12].

The EBU stresses the importance of developing programs for different age groups, especially from 3 to 6 years, 6 to 12 years, 12 to 18 years and 18 to 34 years. This is one of the weaknesses of RTVE, since its audience is mostly made up of older adults. The network's children's channel, *Clan TV*, is a leader in the groups of 4 to 9 years old and 10 to 12 years old.

To justify its mandate, public television must offer something different from private broadcasters. In that sense, we can mention the donation of equipment RTVE has made to non-profit organizations and the many awards it receives each year. Another differentiating factor is that it is a producer and promoter of culture. According to Kantar Media's yearbook of audiences, more than 57% of the programs of TVE's second channel TVE are cultural programs.

4.3. Recommendations for the future

The financial aspect is one of the weakest areas of the public corporation. RTVE managed to reduce its debt in 2007, which was taken over by the State; in the following years it re-created that debt, but not at the same level. Moreover, the budgetary adjustment has had an impact on the audiovisual production sector, which has lived with the concern that RTVE will not invest as much money in programs as it has done in the past [VAN 12].

Due to the ongoing reforms in RTVE and especially the economic context that has surrounded the Corporation, there has not been a President of the Corporation that has lasted more than 2 years in office. The previous presidents since the last reforms have been Luis Fernández (2007–2009), Alberto Oliart (November 2009–July 2011), Leopoldo González-Echenique (June 2012–September 2014) and Jose Antonio Sanchez (since October 2014) [OJE 11].

The recommendations to be suggested to TVE are to continue investing in the production of quality content different from that of private television and to maintain the work started in online innovation and development. There are other aspects that should be improved, for example, being a source of reliable, plural and independent information, attracting young audiences and establishing a stable corporate strategy. If there should be a public broadcaster, it must have stable funding, guaranteed editorial independence and respond to an efficient management.

4.4. Bibliography

[ALM 10] ALMIRÓN N., "The Regulation of Public Broadcasters' News Coverage of Political Actors in Ten European Union Countries", *Comunicación y Sociedad*, vol. XXIII, no. 1, pp. 205–236, 2010.

[BUS 89] BUSTAMANTE E., "TV and public service in Spain: a difficult encounter", *Media, Culture and Society*, vol. 11, no. 1, pp. 67–87, 1989.

[BUS 07] BUSTAMANTE E., *Storia della radio e della televisione in Spagna, 1939–2007: il lato debole della democrazia*, Rai-ERI, Roma, 2007.

[CAL 08] CALLEJO J., "La presencia de la sociedad civil en la televisión pública", *Zer: Revista de Estudios de Comunicación*, vol. 13, no. 24, pp. 29–51, 2008.

[CAM 13] CAMPOS-FREIRE F., "El futuro de la TV europea es híbrido, convergente y cada vez menos público", *Revista Latina de Comunicación Social*, vol. 68, pp. 89–118, 2013.

[COS 12] COSTA C., PIÑEIRO T., "Nuevas narrativas audiovisuales: multiplataforma, crossmedia y transmedia. El caso de Águila Roja (RTVE)", *Icono14*, vol. 10, no. 2, pp. 6–28, 2012.

[EFE 13] EFE, "Bruselas retira su denuncia contra la tasa a las 'telecos' para financiar RTVE", available at: http://www.rtve.es/noticias/20130718/bruselas-retira-denuncia-contra-tasa-operadores-telecomunicaciones-para-financiar-rtve/717882.shtml, 18 July 2013.

[ELD 14] ELDIARIO.es, "Los trabajadores de TVE convocan una jornada de protesta contra la censura y la manipulación", *El Diario.es*, available at: http://www.eldiario.es/sociedad/asamblea-trabajadores-RTVE, 16 September 2014.

[FER 14] FERNÁNDEZ E., PEÑA R., "Plante de los periodistas de TVE por la revolución en los Informativos", *El Mundo.es* available at: http://www.elmundo.es/television/2014/11/28/5478a583ca4741166b8b456b.html, 28 November 2014.

[HER 11] HERRERO M., FRAGO M., MEDINA M., "Inmigración y medios audiovisuales en el mercado español", *Revista de Comunicación*, vol. 10, pp. 25–50, 2011.

[IES 09] IESE, Estudio anual tendencias del sector audiovisual español. Resumen ejecutivo, IESE Business School, SP-SP, Universidad de Navarra, Barcelona, 2009.

[MAN 11] MANFREDI J.L., "Escenarios y retos de la televisión pública en España", *adComunica. Revista Científica de Estrategias, Tendencias e Innovación en Comunicación*, no. 1, pp. 49–62, 2011.

[MED 10] MEDINA M., OJER T., "The New Spanish Public Service Broadcasting Model", *Comunicación y Sociedad*, vol. XXIII, no. 2, pp. 329–359, 2010.

[OJE 11] OJER T., "El papel del gobierno corporativo en la buena gestión de las empresas públicas de televisión: el caso de RTVE", *Anàlisi*, no. 43, pp. 59–71, 2011.

[PAL 05] PALACIO M., "Early Spanish television and the paradoxes of a dictator general", *Historical Journal of Film, Radio & Television*, vol. 25, no. 4, pp. 599–617, 2005.

[PER 65] PÉREZ M., EZCURRA L., *La televisión*, Editora nacional, Madrid, 1965.

[RIC 94] RICHERI G., *La transición de la televisión: análisis del audiovisual como empresa de comunicación*, Bosh, Barcelona, 1994.

[RTV 15] RTVE, Annual Report, available at: http://www.rtve.es/contenidos/, 2015.

[SPA 06] SPANISH COURT OF AUDITORS, Gestión de Personal del Grupo Radio Televisión Española, 2004–2006, Report, 2006.

[VAN 12] VANITATIS, "Las productoras acusan a TVE de llevarlas a la "parálisis" por los recortes de ficción", available at: http://www.vanitatis.elconfidencial.com/cine-tv/2012/03/23/las-productoras-acusan-a-tve-de-llevarlas-a-la-paralisis-por-los-recortes-de-ficcion-18139-denuncia-manipulacion_0_303770554.html, 23 March 2012.

5

The Battle for Audiences

In this chapter, we will address the evolution experienced by the industry in the battle for the audience. This will require, first, to review the economic rationale for the conquest of time in traditional TV networks. We will point out some ideas about digital audiences and appropriate measurement systems that come as a result of digital development. Finally, we will note the need to advance in the knowledge of the audience from a more qualitative perspective. We no longer deal with the battle for the time of the viewers, but the battle for their attention, for engagement with them.

5.1. The battle for time in the media market

The media market, in particular the television market, is actually a market of time: players in this market compete for who attains the time and attention of the public out of all the entertainment options offered. As Nieto [NIE 00] points out, it is possible to say that time can be the subject of a relation (commercial, informational, cultural, etc.), which means that time can be seen as a commodity or as something that can be the subject of deals or contracts, which can be bargained, or that we can "hire" a person's time.

In this sense, it is possible to speak of time and attention as the scarce goods traded in the entertainment market: although it is possible to choose from many entertainment options, the available time and money to consume are limited, as well as the attention that can be dedicated to the offered products.

Chapter written by Mónica HERRERO.

Television activity is mostly financed by income received directly from advertisers, but not as compensation for television products. Advertisers do not buy programs; they buy airtime to insert their ads. And not any time either, but airtime that corresponds to programs whose audience is right for their interests. In this sense, commercial television does not produce programs, but audiences to be sold to advertisers [OWE 92]. This activity becomes one of the fundamentals of commercial television and the logic behind television programing.

With this in mind, it is interesting to identify the kind of information that helps advertisers make their purchasing decisions for broadcasting time. Daily audience measurements provide information about the preferences of viewers and trigger economic processes dealing with high amounts of money. It is possible to question the validity of the information these measurements provide, but the industry has agreed to accept them as a reliable measure, heedless of the intricacies of the activity of watching TV. As an activity that happens in the private sphere, it is difficult to control from the outside the quality of the attention given to the content and the influence of advertisements in purchase decisions.

Thus, the activity of commercial television is dominated by uncertainty, for both the TV channels and the advertisers. For the channels, because their activity of programming will always face the uncertainty of the audience figures they need to reach, advertisers have to invest money before they know the audience who is going to watch the programs in which their ads are inserted. Besides, there are many conditions that can alter the estimates they are using to guide themselves. However, the industry has assumed that uncertainty and has accepted the audiences' data supplied by audiometric companies.

The consumption of commercial television is associated with the programming and the time limits imposed by the chain. An established programming schedule allows programs to accumulate audiences who have acquired the habit of sitting in front of the TV at scheduled times, enabling fixed advertising rates. The prediction of audiences for the different slots and blocks in television programming makes up the core of the business. That is why traditional television maintains the concept of "prime time", based on the consumers' habits and organization of time.

Knowing the television audience includes knowing their tastes and preferences. Audience figures do not report the degree of public satisfaction which, to some extent, has been sufficient for the business of television companies. As explained by Sánchez-Tabernero [SÁN 08, p. 115], "the assessment that the public makes about the information or entertainment products does not affect the income of media companies: users pay the same and produce the same advertising revenue regardless of their idea that a specific chosen product has been excellent, regular, or just the least bad one in the market". What matters to them are the figures of viewers.

This apparent disconnection between valuation and consumption came from the inability to show preferences on commercial television. The absence of a purchase price prevented viewers from really showing their interest and preference over contents, and therefore kept the television market from efficiently providing goods that could be desired. This is why pay-TV has been considered the most efficient. Laguna de Paz [LAG 94] stated that from an economic point of view, pay-TV is the most efficient, because it excludes the character of public good of television, creating a direct relationship between the supplier and the consumer.

Audience figures, however, do indicate the time dedicated by the different profiles to different programs, somehow showing preferences for one genre or the other and helping us advance in the knowledge of those preferences. In many countries, generalist TV channels are still leaders and this seems to be due to two strengths: the variety offered in their programming and the relationship the television channels maintain with their usual public.

Some audience data in Spain allows us to prove this reality. The average daily consumption of television has increased every year. As an example, we went from 227 min of TV viewing in 2008 to 246 min in 2012. However, for the television companies, it is of particular interest to pay attention to different age groups: the commercial target (between 25 and 44 year olds) is the most attractive to advertisers; the younger groups are also very interesting as they can help project the future use of television. Nevertheless, the population group that consumed the most in 2011 is the one of 65 year olds and over (317 min in 2008 to 330 min in 2011) while the group that watched TV the least (144 min in 2008 and 148 min in 2011) is the one of 13–24 year olds. The commercial target, which would be between 25 and 44 year olds, consumed an average of 201 min a day in 2008 and 209 min per day in 2011, not a petty figure. These data may reflect an obvious reality:

people over 65 years of age have more free time. They also have developed the habit of free and programmed audiovisual consumption; a habit developed thanks to the commercial television organized programming that simplifies viewing by saving the consumer the need to search too much for content.

Moreover, the relationship of the viewer with the TV channel, either by habit or brand engagement, still has its weight in guiding consumption [HER 15]. At the moment, audiences continue to maintain a traditional relationship with broadcasters, and it will likely remain so while the older age groups maintain the weight they currently have on the average audience figures. The search for varied contents and the reference to a brand identity leads us to think that the audience still goes to traditional TV channels in search of content management to face the huge offer available. The system of the traditional linear supply, with programming organized in blocks, facilitates variety and passive consumption at times of rest.

Traditional broadcasters play an important role as prescribers of content, as places where the audiences go in search of the content they are interested in the new digital context. In this sense, we can distinguish between linear services, where the service provider decides when a specific program is transmitted and establishes the program schedule, and nonlinear services, in which it is the user who decides the time to view a specific program based on a choice of content selected by the service provider.

Therefore, regardless of the chosen service provider, platform or device, content is becoming increasingly more important as the key of the television business. If commercial channels respond to these challenges with innovation and determination, they will assume an undisputed leadership in the coming years.

5.2. New measurement systems

As we have seen in previous paragraphs, broadcasters need to know their audience in order to offer the best content in the most appropriate schedule and reach the most valuable audiences [BAL 11, BER 13] so that they can design the most profitable programming strategy. Therefore, offering reliable audience data is fundamental for the commercial television market [WEB 14]. If the offer of online video content looks for commercial revenues, audience data are required. However, the measurement in the digital era is not easy, because the consumption of video takes place across screens and time.

However, different systems to measure audience of online video content have already been developed. One possibility is the Audio Matching System (AMS). It permits the identification of content with audio data. The most recent systems assign an identifier or tag (an audio watermark) to each content that is matched with the channels broadcasting in order to identify the content viewed [UNI 09, KAN 12, NAF 16]. In the United Kingdom, *BARB* is already testing the use of identifiers to report "consumption of online TV content in TV players controlled by the broadcasters" [BAR 15]. The project is called *Dovetail* and it is run by with *Kantar Media* and *Nielsen*. The data are based on tags and a combination of Internet census data and demographics from the panel [BAR 15, POR 15]. This is called a hybrid measurement. *Kantar Media* and *comScore* are also testing this combination of data and the use of tags [GOO 15].

Portable people meters (PPM) can also be a possibility [POR 07]. The *Ipsos's MediaCell* captures exposure to encoded audio signals by just uploading software into the panelists' smart phones. The advantage of the cell is that people take the meter everywhere without requiring an extra device, as previous PPM [POR 07]. *Nielsen* is also testing mobile/web apps to replace paper diaries for US local market television measurement [SHA 13]. The problem is that the audience figures generated by traditional meters and newer passive devices can vary, sometimes considerably [GRE 11].

Kantar Media had also been testing a PC meter system to measure online activity, installing software among some of their Spanish television meter panelists [NAF 14]. It is the Kantar's Virtual Meter technology. However, *Kantar* has moved from this project to the hybrid measurement with *comScore*, combining census data from Internet plus demographics from the panel of users [COM 15, NAF 16].

In brief, the measurement of online video consumption across multiple screens is still a challenge. The main market research companies measuring audience are still trying to find the best system. At the same time, advertisers are focusing on the proposals, because they do not want new metrics. Advertisers promote projects using the existing metrics for cross-platform audience analysis [ESP 13]. Therefore, achieving a consensus among participants in the media market is the key.

5.3. Toward qualitative research

The relationship between content and audiences in the new digital media market context signposts how the attention of the public may be engaged and how audience satisfaction may be measured, phenomena that are especially evident in the case of audiovisual media content.

As occurred with the introduction of cable television, the availability of broadband television signal broadcasting presents telecommunications operators with new business opportunities, as well as other suppliers outside the TV industry who have seized the chance to expand their business into the booming audiovisual sector. Television networks are no longer the only providers of audiovisual information and entertainment contents. Users have begun to play an increasingly significant role in content development – see the Web 2.0 model, for instance [DAV 97, SHA 99]. User-generated or homemade audiovisual content receives high viewer ratings or number of hits on global online platforms such as *YouTube*.

New technologies have given individual citizens the status of co-editors in shaping public opinion via blogs, forums, online communities and instant messaging; neither the traditional media nor politicians can afford to turn a blind eye to this shifting influence.

The potential for personalization enabled by new technologies may lead to the paradox described by Carlson [CAR 06]. While the new media grants users the power to enjoy media content when and as they choose, and to avoid advertising, such technologies also enable the collection of valuable audience-related data of great interest to content providers and advertisers, such as demographic information and TV viewer preferences. In this regard, users may assert greater control over the programming schedule, but television companies, networks and advertisers acquire parallel access to individual user data, thus facilitating a more targeted form of advertising.

Digital media audiences tend to be more passionate and engaged: they do not simply enjoy what they like by themselves, they share it with others [NEW 12]; hence, communities based on shared interests, tastes and knowledge emerge. At the same time, the evaluation of contents rests on popularity – in general, a spontaneous form of popularity – which content producers may find difficult to either plan for or predict.

As television programs develop to allow more interactions between them and their audiences, more interactive forms of reality television encouraging heightened audience activity will be developed. Television series, especially soap operas, are also likely to engage greatly with the audience. The series go beyond their broadcast time not only in reference to the real world of the individual but also in the relationships with other followers. Some media products, such as drama, seem more compatible with the value of SNS and consumer engagement [DIE 10].

Media audiences are defined on a scale of quantitative parameters that advertisers use to budget their investment, based on viewer rating figures for television. However, content producers are aware that users have a variety of tastes, feelings and critical perspectives with regard to different media products. The critical attitude among the audience is especially acute in relation to content for which they have to pay. Audiences that pay, known in commercial circles as subscribers, tend to pay only for content they regard as indispensable [HER 03, PIC 11].

Pay-per-view media makes a wide range of consumer options available, and fosters user evaluation and personalization of the content on offer: users may pay to view the basic package, select specific content or pay to access exclusive content. Pay-per-view options grant the subscriber a decisive role in establishing which content ought to be offered and which withdrawn on the basis that few or no users express an interest in them. The pay-per-view model facilitates an almost perfect match between supply and demand, taking into account the tastes and needs of subscribers.

To engage and hold the audience's attention requires a refined understanding of who the audience is. Hence, audience research is a key aspect of content management. Callejo [CAL 01] listed four types of audience research: *diffusion studies*, which draw on audience ratings figures used for commercial purposes in the sale of advertising space and are based on the viewer‘s use of television; *structure studies*, which are carried out by marketing departments by means of surveys, interviews and focus groups organized to assess the product, so as to shape executive action in the market; *reception studies*, an academic-led approach to defining the audience and, as a consequence, the society in which contents are imbricated; *integration studies*, whose purpose is to involve the audience in media products, and which are promoted by the production companies themselves.

The meeting point with consumers, which is a key location for companies in the service industry, is not a physical space for media users, where direct interaction may take place. Rather, ratings figures amount to an implicit, indirect form of contact, wherein the role of users tends to be minimized.

Westerlund [WES 11] highlighted two aspects of the new digital business model: the revenue paradigm and customer-driven strategy. Both approaches foreground the role of the audience. The transformation of business practice brought about by new technologies has prompted a shift from something-for-everyone to everything-for-selected media consumers, that is, specialized and personalized content that responds to consumer profiles. The internal structure of media companies has similarly been transformed by this radical change, leading to a focus on market research, as well as on building organizational capabilities and operational skills. The company- or competitor-driven strategy, which had shaped media company operations for many years, has been replaced by the customer-driven strategy. This shift also requires that business organizations be more versatile and adaptive [SÁN 95].

In addition to allowing users to choose their favorite programs or channels, the type of interactivity enabled by new technologies fosters what Bordewicjk and Kaam [BOL 86] referred to as interactivity of register, which represents the media's capacity to register user information and adapt or respond to user needs. Thus, television has become a more refined advertising vehicle than any other media, which does not facilitate such interactivity. At the same time, combining the register approach with ratings figures may enhance audience satisfaction with television programs. Despite their limitations, the media may become personalized services, whereby the audience plays a direct role in content creation. Prahalad and Ramaswamy [PRA 00] draw a distinction between *customization*, a production strategy whose purpose is to meet the needs of a particular client, and *personalization*, in which the client is a co-creator or the content he or she uses. The latter strategy is especially important in the entertainment industry.

Digital media foster a more active attitude to contents among users. By means of technologies integrated into the platform, rather than external devices such as mobile telephones or the Internet, the audience takes part in content creation and evaluation. As Albarran [ALB 10, p. 32] averred, "audience members access information when they want, where they want, and how they want".

Although the Internet is still a new form of media and its audience measurement tools are not yet very well developed, it is clear that you can not only quantify the number of users or the amount of time spent on a given website but where users come from and what they like may also be captured. Thus, the Internet provides more detailed knowledge of the audience. Therefore, the goal of online media is to match the product to audience interests, rather than merely to feature in the top 10 [ALB 10].

The audience does not simply consume contents: implicitly or explicitly, the public acknowledges the value of such contents insofar as they add to their knowledge and/or meet their expectations.

If contents succeed in holding the public's attention, then audience satisfaction increases. However, establishing a yardstick against which such satisfaction may be measured is not a straightforward task in media management. In 1995, the annual report issued by AEDEMO (the Spanish Association of Market, Marketing and Opinion Research) included an assessment scale devised to optimize program scheduling in audience satisfaction terms, based on viewer predictions in relation to programming [PAS 95]. To ensure that these predictions are accurate, the assumption is that every individual is watching what he or she likes in any given moment – the so-called *coefficient of coherence*. Further research disclosed that this was not in fact the case, and researchers began to ask why viewers do not always watch the program they might like most. The reasons for this apparent contradiction are many and various, and often lie beyond the control of media managers: the consumer choice may depend on someone other than the individual viewer; viewers may be unaware of other, more appealing program options; broadcast schedules may prevent viewers from watching the programs they most like and loyalty to a particular program may lead viewers to ignore more highly rated programs scheduled for broadcast at the same time.

Once these divergent factors have been controlled for, program reception may be used by television channels as part of the scheduling decision-making process. At the same time, given that they may depend on subjective criteria that cannot be easily quantified, program audience figures comprise one of the least predictable forms of data in the media sector.

Frank and Greenberg [FRA 80] carried out a series of inquiries exploring the interrelationship of audience attention, program types and the reasons why viewers watch different programs. Among the significant potential factors in this regard are social stimulus – that is, sharing ideas with others; the desire to retain a certain social status and/or to impress others; to assert leadership; to escape one's problems; to build closer family bonds; to understand others; entertainment; to improve one's knowledge; an intellectual desire to learn. As Frank and Greenberg pointed out, this research facilitates program scheduling based on audience interests, program promotion and the evaluation of existing shows.

Attracting and holding the audience's attention marks success in understanding what audiences need, their likes and tastes. Meeting audience needs and interests involves fulfilling their wishes, thus raising their spirits and prompting them to share what they like in others. In the new digital media economy, sharing is both a sign and measure of satisfaction. Satisfaction can not only be assessed in terms of audience ratings, but can also be evaluated with reference to the number of individuals who talk about the content; how many of them view the content repeatedly; how many download or record contents for later viewing; how many register as fans or access other media to find out more about given content and how many are prepared to pay for related products [MED 09].

5.4. Bibliography

[ALB 10] ALBARRAN A., *The Transformation of the Media and Communication Industries*, Media Market Monographs, Pamplona, 2010.

[BAL 11] BALNAVES M., O'REAGAN T., GOLDSMITH B., *Rating the Audience: The Business of Media*, Bloomsbury Academic, London, 2011.

[BAR 15] BARB, The TV Player Report – *FAQs*, available at: http://www.barb.co.uk/tv-player-report/release/383, 8 December 2015

[BER 13] BERNÉ C. *et al.*, "Understanding the consumption of television programming: Development and validation of a structural model for quality, satisfaction and audience behaviour", *International Journal of Marketing Studies*, vol. 5, no. 1, pp. 142–156, 2013.

[BOR 86] BORDEWIJK J.L., VAN KAAM B., "Towards a New Classification of Tele-Information Services", *InterMedia*, vol. 14, no. 1, pp. 16–21, 1986.

[CAL 01] CALLEJO J., *Investigar las audiencias. Un análisis cualitativo,* Paidós, Barcelona, 2001.

[CAR 06] CARLSON M., "Tapping into TiVo: Digital video recorders and the transition from schedules to surveillance in television", *New Media & Society*, vol. 8, no. 2, pp. 97–115, 2006.

[COM 15] COMSCORE, "ComScore and Kantar Announce Strategic Global Partnership to Accelerate Cross-Media Audience and Campaign Measurement". *Press Release*, available at: https://www.comscore.com/Insights/Press-Releases/2015/2/comScore-and-Kantar-Announce-Strategic-Global-Partnership, 12 February 2015.

[DAV 97] DAVIDSON J., REES-MOGG W., *The Sovereign Individual: How to Survive and Thrive During the Collapse of the Welfare State*, Simon and Schuster, New York, 1997.

[DIE 10] DIEGO P., HERRERO M., "Desarrollo de series online producidas por el usuario final: el caso del videoblog de ficción", *Palabra Clave*, vol. 13, pp. 325–336, 2010.

[ESP 13] ESPN, "ESPN's 5-Platform Measurement Initiative with Arbitron and comScore", *Presentation at the CIMM 2nd Annual Cross-Platform Video Measurement Summit*, available at: http://cimmusorg.startlogic.com, 2013.

[FRA 80] FRANK R.E., GREENBERG M., *The Public's Use of Television: Who Watches and Why*, Sage, Beverly Hills, 1980.

[GOO 15] GOODE P., *Cross Media Kantar Media comScore*, Presentation at the 10° Encuentro AIMC, Madrid, 20 November 2015.

[GRE 11] GREEN A., "Understanding television audiences", Warc Best Practice, available at: http://www.ipsos.com, September 2011.

[HER 03] HERRERO M., *Programming and Direct Viewer Payment for Television*, Media Markets Monogaphs, Pamplona, 2003.

[HER 15] HERRERO M., URGELLÉS A., "La audiencia y hábitos de consumo", in MEDINA M. (ed.), *La audiencia en la era digital*, Editorial Fragua, pp. 67–85, 2015.

[KAN 12] KANTAR, "RapidMeter: A simple innovation bringing audience measurement to new markets", *Kantar News and Views*, available at: http://www.kantarmedia-audiences.com, 11 July 2012.

[LAG 94] LAGUNA DE PAZ J.C., *Régimen jurídico de la televisión privada*, Marcial Pons Ediciones Jurídicas, Madrid, 1994.

[MED 09] MEDINA M., *Creating, Producing and Selling TV Shows. The Case of the Most Popular Dramedies in Spain*, Media xxi, Lisbon, 2009.

[NAF 14] NAFRÍA E., "Del TAM al VAM", Presentación en el *30° Seminario TV 2014 de AEDEMO*, February 2014.

[NAF 16] NAFRÍA E., "Con el acuerdo entre Kantar Media y ComScore, España se coloca en la línea de salida hacia el *total view*", *Investigación y Marketing*, vol. 130, pp. 14–17, 2016.

[NEW 12] NEWMAN M., "Free TV: file-sharing and the value of television", *Television and New Media*, vol. 13, no. 6, pp. 463–479, 2012.

[NIE 00] NIETO A., *Time and the Information Market: The Case of Spain*, Eunsa, Pamplona, 2000.

[OWE 92] OWEN B.M., WILDMAN S.S., *Video Economics*, Harvard University Press, Cambridge, 1992.

[PAS 95] PASCUAL J., NAVIA C., "Una aproximación a la predicción de audiencias: panel de valoración de programas", *Report 11° Seminario de Audiencia de Televisión,69° Seminario de AEDEMO*, pp. 201–218, 1995.

[PIC 11] PICARD R.G., *Mapping Digital Media: Digitization and Media Business Models*, Open Society Institute, 2011.

[POR 07] PORTILLA I., "La medición de audiencias en televisión", *La televisión en España, Informe 2007*, Ediciones Deusto, Barcelona, pp. 269–283, 2007.

[POR 15] PORTILLA I., "Television audience measurement: proposals of the industry in the era of digitalization", *Trípodos*, vol. 36, pp. 75–92, 2015.

[PRA 00] PRAHALAD C.K., RAMASWAMY V., "Co-opting customer competence", *Harvard Business Review*, pp. 79–87, January–February 2000 Issue, 2000.

[SÁN 95] SÁNCHEZ R., "Strategic flexibility in product competition", *Strategic Management Journal,* vol. 16, pp. 135–159, 1995.

[SÁN 08] SÁNCHEZ-TABERNERO A., *Los contenidos de los medios de comunicación: calidad, rentabilidad y competencia*, Deusto, Barcelona, 2008.

[SHA 99] SHAPIRO A., *The Control Revolution*, Public Affairs, New York, NY, 1999.

[SHA 13] SHAGRIN C., LINK M., "Reinventing TV Diary Measurement – Update from CIMM-Nielsen Innovation Committee", *Presentation at the CIMM 2nd Annual Cross-Platform Video Measurement Summit*, available at: http:// cimmusorg.startlogic.com, 2013.

[UNI 09] UNITAM, Unitam Technology. UNITAM for Sweden, available at: www.mms.se, 2009.

[WEB 14] WEBSTER J.G., *The Marketplace of Attention: How Audiences Take Shape in a Digital Age*, MIT Press, Cambridge, 2014.

[WES 11] WESTERLUND M. *et al.*, *Insights into the dynamics of business models in the media industry*, Laurea Publications, Helsinki, 2011.

6

Entertaining Power: Quality of Content

Quality is a relative value that refers to a paradigm of excellence. This is what allows us to speak of high or low quality when a product or service approaches or moves away from that paradigm when valued.

TV quality becomes an issue with the privatization of the sector in the late 1980s. The entry of private capital and increased competition influenced the discussion on an issue that until then was assumed within the concept of public service. Quality of TV content was irrelevant in a context of monopoly, where supply was limited to public channels. Despite the open debate, there is still no consensus on what quality television is and what the objective scales and methods to assess the quality of television programs effectively are [PIC 00].

The discussion of quality in TV offers a diversity of opinions depending on the area from which you approach it and there is no agreement between the parties involved. Various actors – legislative and policy agents, professionals, academics, agencies and audiovisual commissions, entrepreneurs, managers, critics and spectators – agree on some aspects but differ about others.

Not clear in the theory, commercial television and practice has also abandoned this useless debate, in a context where raw profitability comes first. The priority of television companies is to increase advertising revenues, which is why this debate is often settled within many companies by identifying quality with number of viewers.

Chapter written by Mercedes MEDINA.

Moreover, the European perspective also differs from the American perspective, partly due to the different development and structure of television on both continents. Even within Europe, different opinions a rise. Certainly, in professional, academic and political forums, the term quality is constantly referred to as an explanation of activities and goals to be achieved.

To define some guidelines for quality in entertainment and fiction is more difficult than defining guidelines for quality in information [COS 05, VON 10]. Although caution always plays a key role in decisions, in the case of information, truth and duty are more evident than the truth and duty of entertainment and fictional content. Quite often entertainment is confused with rest and considers that television does its job simply by filling the time that viewers do not spend working. The real concept of entertainment goes beyond mere rest.

According to the dictionary, to entertain means to recreate the spirit. Therefore, it is a type of activity that requires a response, that is, the receiver of the action needs to feel entertained [GON 99]. Consequently, it is essential to think of the recipient throughout the whole process of ideation, production, marketing and television programming. Knowing the recipients leads to a better chance of success.

Taking the meaning of entertainment as something that elevates the spirit, it is undeniable that the ingredients needed for the task are the result of intelligence and the will of whoever produces and whoever receives the content. When these elements are missing, television programs do not really entertain, they bore and do not facilitate the rest that is necessary in leisure time.

Leisure time is usually *free* time, which will be more free if it is well used, and it will be well used as long as the contents consumed meet the needs of information and entertainment from whom demand them and provide intellectual wealth. The content of the media can entertain people in a way that makes spent time, useful time. People who have intelligent content at hand are more likely to make smart choices. People blunted by feelings and passions will be conditioned by more volatile sensitive and difficult-to-master stimuli.

Sustaining that the primary purpose of television is to entertain does not mean any demerit to the medium [SHU 03]. On the contrary, entertaining is one of the most difficult activities, and doing it successfully is an art that requires wisdom and deep understanding of the concerns, interests and motives of the recipients, especially when we have to entertain a wide audience through contents as diverse as series, contests, news or documentaries.

The failure of some programs reveals a certain weariness of the audience in the face of the lack of variety of programming. This situation of fatigue calls for new ideas in a market where duplication and easy search for success seem to have an end. The need for new ideas and formats becomes urgent in the current environment of multiplication of content offerings.

Despite these theoretical premises and this lack of consensus, quality television demands the commitment of wanting it as much as the necessary market structure. In the end, quality is an expression of a job well done and that is what users demand, even implicitly. Any work entails responsibility and this can only happen in an environment of freedom.

To provide some operational guidelines, it seems necessary to propose a definition of quality that can be applied in business and commerce. In the next section, we will review, from a critical perspective, the different approaches that guide the quality parameters of audiovisual content and some measures that serve to gauge the degree of quality. We will also summarize the changes that the television sector has seen in the late 1990s. Some question whether the new possibilities of distribution and consumption will favor the inclusion of quality criteria in the supply of audiovisual content. Finally, the quality criteria according to technical, thematic and formal aspects will be presented.

6.1. Defending media content quality

Quality in business is all about adaptation to strategic objectives, to the budget of the company and to the public's needs, by offering different, original, fresh and relevant products [OTS 08].

In the field of media, the key question about quality that entrepreneurs have to answer is what content they want to offer according to the identity and mission that distinguishes their company from others concurring in the market. The identity and mission must correspond to the implied and unsatisfied demands of the public, and must be known and shared by the public, employees and shareholders [DIE 11]. Furthermore, in its strategic planning, they must consider the means available, the resources they need to acquire and how much they want to risk to acquire them. Finally, we must not forget the competitive environment in which they operate because differentiation is essential as a competitive advantage.

In this regard, the key points for effective content management and quality content will be the public, competition, productivity of resources and consistency with corporate identity [COR 99].

The potentialities of television are undeniable. The power of the image, movement and sound has a greater influence and impact than any other medium. The scope and coverage, the graphic creativity, expressive richness and ease of use are traits that are sometimes forgotten by imitating other formats and other media.

Quality is very important when building a sustainable brand. It gives a reputation to the brand that facilitates trade, public loyalty, stability and cohesion of the workforce. Quality is a distinguishing feature, engine of business, an intangible value that should always be present.

At the origin of a business in any sector, the public implicitly demands quality products and, in the case of media, the sender or publisher promises their best efforts to get it. This puts the basis for the business relationship between the media and the public. This relationship seems dimmed in the television business, because it is mediated by gratuity, but still, there must be a commitment to quality as the foundation of trust.

On the contrary, it can consolidate the business in the long term and from this position sustained over time, can grow into other sectors, expand activities and spread to other countries. Quality also helps create barriers to entry because quality content becomes difficult to imitate. This is a disadvantage for content like tabloids or pornography, easily imitable and not as good for competitive differentiation. A strategy based on quality as differentiation calls for continuous improvement, which also helps keep employees motivated and satisfied with their good work.

6.2. Perspectives of quality on TV

There are different perspectives that provide complementary visions about the quality of audiovisual content [ISH 96, GUT 00, MED 06]. Academic, administrative and pragmatic perspectives can be summarized into eight approaches: (1) quality and public service; (2) quality and diversity; (3) quality and promotion of European culture; (4) quality and freedom of expression; (5) deontological perspective; (6) cultural studies; (7) applied theory perspective and (8) professional practice.

These approaches are not mutually exclusive. The first two refer especially to the concept of public service television, which are entitled to protection as long as they provide quality service. The following two are attached to the idiosyncrasies of Europe and the United States, respectively. Deontological perspective refers to the ethical principles underlying professional activity. There are two approaches to the subject of the academic field. The first responds to the critical school whose theoretical foundation is in Marxist theory, and the second is moderate school and seeks an understanding of businesses and markets from an applied theory approach. Finally, the point of view of professional practice can contribute to value these approaches according to commercial efficiency.

Quality is generally identified with adaptation to the intended purposes. The origin of television in Europe explains that quality is identified with the functions of public service. The public interest is identified with those shared by the majority of the population of a country, regardless of its cost. This theory defends the existence of the public broadcasters as guarantors of quality and general interest.

The White Paper, *The Future of the BBC. Service the nation competing world-wide* (1994), points out 10 of the objectives that the government believes that the BBC must comply: (1) prioritize audiences; (2) promote diversity; (3) reflect national identity; (4) enrich the cultural heritage; (5) achieving quality standards, which include creativity, originality, technical quality and production; (6) maintain editorial independence; (7) make it accessible to all sectors of society; (8) provide opportunities for new professionals and promote professional development; (9) respond to the interests of taxpayers trying to satisfy their information needs and (10) ensure transparent accounting of its activities.

The United States Federal Communication Commission (FCC) summed up in 14 points what is *not* public interest [HUD 87]. Among other features, this description highlights defamation, racial or religious intolerance, obscenity, etc. Although it is a negative approach, the American version of quality as general interest provides an important aspect, that is, private companies can provide services of general interest. It is not an exclusive competence of the public sector.

The French concept of public interest is closely related to the promotion and expansion of French culture [GRE 72]. The artistic value of programs prevails over entertainment and it must be producer for all audiences.

In this context, the quality values are identified with coverage of the interests of the entire population without discrimination of any group. In Spain, the General Audiovisual Communication Law 17/2010 considers that public broadcasting service is an essential service of general economic interest whose mission is to disseminate content to promote the principles and constitutional values, contribute to the formation of a plural public opinion, publicize the cultural and linguistic diversity of Spain, spread knowledge and arts, with special emphasis on the promotion of audiovisual culture and cater to those citizens and social groups that are not recipients of the generalist programming.

The mode of theoretically defining and measuring quality television is complex. From the administrative and legislative level, it is habitual to use *diversity* as a measure of quality. Diversity ensures creativity, ingenuity, novelty, lack of repetition or excess foreign programs or programs of the same genre, and guarantees the existence of programs to suit all tastes and interests, ages, social classes and groups with educational training and different culture.

The criterion of diversity is a practical way to combine the dimension of public interest with freedom of expression, but by itself is not effective for assessing quality in television. Diversity aims to ensure quality when some genres considered quality are present in the offer, such as cultural programs, documentaries or local content. Thematic specialization can give companies a way to differentiate from competitors while satisfying the viewers and meeting their interests – especially in a context of abundance of offerings, which would reduce diversity, but not quality.

It also allows differentiating television genres according to their purpose, origin and type of production, resources employed, the mode of narration, the audience they are targeting, etc. The typology of programs helps distinguish quality standards according to each genre and its purposes, and avoids the assessment of all programs with the same criteria. For example, drama should create suspense, play with surprise and generate moments of tension; news programs aim to eliminate uncertainties, communicate facts about reality and infuse serenity to citizens.

In order to theoretically support the promotion of *European culture*, the designers of European audiovisual policy were based on the idea of the existence of a European cultural unity, substratum of the diversity of the various European national cultures. These common roots lie in the historical origins of Europe: the expansion of the Roman Empire and the spread of Christianity. Thanks to these cultures, "European values" are identified with the freedom- and respect-derived human dignity based on a transcendent vision of man.

With the assumption of the existence of a Western European culture, European policymakers of the late 1980s promoted it as a way to strengthen the union of countries. Television was thought to be the most effective medium to do so because of its influence and reach. Since the acceptance of the European Directive 89/552 *Television without Frontiers*, member countries of the European Union drive and promote all actions that encourage the production and distribution of European audiovisual works and the education of professionals.

Another way to look at this audiovisual policy is economic in nature and it regards the need to strengthen the audiovisual industry in European countries, to compete against the American audiovisual industry. This approach relies on the distrust for North American production and, implicitly, identifies quality with European production. It suggests that an abundant production with worldwide success cannot maintain quality standards, an idea that does not seem justified. Moreover, the current development of the audiovisual industry has resulted in productions involving international teams in which the origin of the production is not clear. Besides, Europe has produced some content formats like Big Brother-type reality shows, which lack some quality criteria.

Often, the European legislator fixates on what could be called "defense of public or cultural values on television", perhaps to protect citizens against possible damage to national identity. To draw the boundaries between the legitimate defense of indigenous values and an attitude of protectionism that borders a *de facto* monopoly is a difficult task. In the latter case, it can endanger the necessary legal and fair competition. The question is asked quite often with rigorous clarity: where does quality protection end and state tutelage of professionals begin?

To be able to comply with Article 10 of the Universal Declaration of Human Rights of 1950 that states that everyone has the right to inform and to be informed, it is necessary that *freedom of expression* is recognized and safeguarded by the competent authorities. However, we must also ask how that freedom of expression is guaranteed and what its limits are.

From the European Union, a defense of competition for pluralism and against corporate concentration is exercised. Concentration is seen as a threat to freedom of expression, because it represents a limit to the number of voices that have access to information, so some sectors of society are in danger of not being represented in the media. To prevent that only a few voices can be heard, limits on media ownership and monopoly control of markets are suggested.

Some important aspects underlined in this approach contribute to shaping quality values. For example, the desire to spread the truth, which needs special interests eliminated in order to preserve the common good; the conception of the audience as people with critical judgment that can think about the different information they receive; the importance of professional criteria over purely economic interests and respect for social values that transcend the information communicated [LEG 96, ALB 92]. In this line of thought, the work carried out in several Latin American universities is worth highlighting. Led by the Catholic University of Chile, they have defined Journalistic Added Value as a method of valuation of journalistic information. Intangible aspects are considered at the time the journalist selects the news, casts a glance on the reality and establishes priorities from it [PEL 06]. Meanwhile, at the Catholic University of Argentina, a list was elaborated of quality standards for journalistic information (CIP). It includes reliability, relevance, interest, proportion, adequacy, transparency, accuracy, clarity, integration and incidence [TER 06]. These values seem easy to apply to information, but not so much to entertainment.

Freedom of expression, in this context, is not an absolute value and must be exercised in compliance with other principles that are above it and give it meaning, such as privacy, security and order, and rule of law. Freedom of expression must be exercised as a means to an end; it is in its nature. To be able to exercise freedom of speech, it is necessary that the required media exist and that the existence of companies for various purposes in the field of a free market economy can be ensured.

Many decisions affecting the activity of television companies have ethical dimensions, because they have consequences for third parties that also leave their mark on whoever decides and executes them. Most of these decisions can be considered from the perspective of quality and from the professional dimension of the media.

In some countries, the ethical principles of the profession have been collected in codes of conduct or manifestos signed by professionals. Although their intention is to ensure the good work and not to provide quality criteria, its guidelines can be used to assess the degree of quality of content. A general note of these codes of conduct is that they seek to promote the presence of positive values in television programs; they brand as unethical content that seems more or less irrational and perverse for civic coexistence. Some rules limit the broadcasting, at certain times, of programs where violence, sex or situations of psychological abnormality inevitably accompany the winner, and where attitudes that have no other support than instinct are presented as "normal". They generally have a protectionist approach toward young audiences, since children are the most vulnerable group and their critical capacity to assess media messages is lower.

In general, they show a preventive attitude toward advertising as a cause of distortions, misunderstandings and deceptions. For example, the resolution of 17 April 1990 on inadmissible advertising in RTVE noted that no advertising be allowed (…) that incites violence and anti-social behavior, appeals to fear or superstition or indirectly encourages abuses, recklessness, negligence or aggressive behavior. Advertising susceptible to incite cruelty and maltreatment of persons and animals or the destruction of cultural or natural assets is also rejected. However, it is not easy to find such strong views on the rest of the programs.

Perhaps, the problem is to bet on rule that set generic prohibiting minimums, and maximums that tolerate and expand easily. Codes are often nothing more than defensive levels of permissiveness. This is why codes of conduct on television are not entirely illuminating, because most of them, at least those published in Western countries, define quality television in negative terms, that is, quality means to exclude sex scenes, morbid curiosity and violence. This exclusive approach does not provide what is necessary to produce quality content guidelines. Furthermore, quality should be an objective of content aimed at children and the rest of the public.

Self-regulation in the area of content is not yet common practice in many countries. Faced with the limitations and injustices of legal systems, subject to the political whims, Audiovisual Organisms in which every professional sector is represented (producers, distributors, customers and viewers) is a reality that has been growing [AZU 09]. The success of these organisms depends on them maintaining political independence and being representative and coordination figures for all the professional and social sectors.

Codes are often criticized for their corporatism and professional inbreeding and for lacking sanctions to give effect to the agreed principles. Notable exceptions are the codes of Canada, Japan and the Philippines, with statements of principles more explicit and detailed than in the other countries [VIL 00].

The study of the effects produced by the hegemonic position of the media is the subject of what is known as *cultural studies*. With a critical approach that is based on the thought of Habermas and the Frankfurt School, the authors applied the theory of class struggle to the media, considering the power of influence of the latter and the limited capacity of the audience to process media messages. They understand that the media broadcast a dominant culture that contributes to building communities of knowledge and relationship to a point of creating reality.

The main objective is not to develop a theory about the quality of contents, but they set off from the premise that media meet their mission successfully when they represent with certainty and balance all social groups and interests, and therefore, its contents will respond to higher levels of quality. Through content analysis, they analyze how different social groups are represented in media messages, paying special attention to racial minorities, feminism and homosexuality.

Even though they do not make a proposal for values of quality, the critical method leads them to denounce the lack of pluralism that exists in the media content, both in information and fiction. However, their concept of *mass* audiences, amorphous and incapable of reacting, becomes obsolete with the advent of new technologies that allow the viewer to choose the content and services they want to enjoy. Moreover, to develop quality guidelines, it is not enough to pay attention only to the recipients, but the vision must extend to producers and distributors [HOG 89].

From a moderate perspective, media economics deals with the analysis of the media from the free market and the strategic direction of the companies. The authors of this school are aligned to a tradition of applied research, by observing reality and analysis of specific cases [MED 98]. Most liberals believe the state should not intervene in the market. The companies serve the needs of the audience with quality services. If they do not comply with this, the market itself will expel them. Others, more moderate, believe that regulation should ensure the common good and avoid dominant positions. In defense of the citizens, they believe that laws should require that content has a minimum of quality in order to be made available to citizens.

The UK's Broadcasting Research Unit (BRU) [BRO 89] suggests a number of professional qualities that contribute to achieving higher standards of quality in television. For example, show society as it is, with its good and bad aspects; invite the audience to relate that external reality with their private lives and teach new perspectives that contribute to enrich knowledge. According to Albers [ALB 96], three notes affect professionals: know their work very well, adapt to new ways of saying and understanding reality and distinguish the relevant aspects from passing trends or issues based on sentimental and simplistic perceptions. For Wolf [WOL 99], the crucial problem of broadcast journalism is to define the criteria of relevance.

The difficulty the communication entrepreneur faces is that it seems that quality is incompatible with economic profitability. From the business perspective, quality relates only to large number of viewers and revenue. Statements like "quality is what the audience demands" or "we offer what audiences want to see" summarize the reasons for the decisions of many managers of television companies, but turn out to be a rhetorical exercise more than an operating principle to pursue quality media. Sometimes, these sayings hide lazy and unimaginative practices [SAN 00].

It is undeniable, from the business dimension, that quality television cannot be identified with content that nobody sees. However, it is also true that many programs now reaching millions of viewers lack some minimum quality standards. Audience is one of the most important elements in the development of content, but serving the audience has to do with meeting their needs, tastes and hobbies; introducing attractive elements; effectively promoting the programs and directing them to the appropriate audience. According to Hillve [HIL 95], audiometric data give an idea of the programs that people like, but not why they like them. However, the development of new technologies to meet the tastes, opinions, habits and willingness to pay of viewers helps the offerers meet their supply to the needs and tastes of audiences and provide quality, for which the viewer is willing to pay.

In all these aspects, there are some elements that make up the quality of the content, but none seem complete. By adding to this reflection, the changes that new technologies have brought to the audiovisual sector will allow us to take into account other systems of quality control of content.

To think about quality content in the current context, it seems necessary to reflect on the changes that digitization has produced and how they have influenced the television business in recent years.

In the digital context, television is no longer a *mass media* and has become a *self-medium*, that is, a personalized media, in the sense that the viewer chooses what to watch, when and where they want [URR 08]. There has been a separation between the emission and consumption, leading to a change of parameters when programming television content.

This selection leads us to distinguish the new values of quality and above all, new measurement systems for such quality. Spontaneous popularity breaks previous economic schemes. Low-budget contents can be the favorite and most viewed by users worldwide. Sharing has become a clear sign of identification with content [SHU 03]. Something being shared is a sign of approval.

Sometimes, these shared content are not created by professionals or within an organization. They are often produced by users, and promoted and distributed by parties outside the authors, who have simply found value in them and want others to enjoy that value.

6.3. Searching for quality

Content is the keystone of the quality of a program. Content is the result of the union of creativity and technical work, personal efforts integrated into an information product ready to be distributed. The making of a TV program is a process in which ideas are woven and taken care of in every stage of production: creative development, production, distribution and marketing. The program is the result of an intellectual structure conformed to satisfy the needs of the recipients, with elements from fiction, non-fiction, various narrative genres and technologies in constant innovation.

The traditional aphorism that says that people watch programs rather than channels acquires a greater force with digital media. TV channels are no longer the only doors to access programs. If content is attractive, the viewers will look for it themselves, regardless of the channel that broadcasts it. The digital economy facilitates the offer of varied content. Therefore, rather than looking for quality in content, viewers look for content of quality. There is no longer a vision about some few genres embodying quality, because quality can be found in all types of programs, as long as the nature and purpose for which they were produced are respected.

For a complete perspective on quality, we must address the three dimensions of media content: technical, thematic and formal.

Frequently, quality in television is related to technical quality only and technical standards are set regarding transmission and reception without interference, accuracy, clarity, timeliness, synchronization between image and sound, balanced use of cameras, variety of angles, adequate lighting and visually appealing images.

Timeliness is very important in processes at any level, especially when interacting with customers or agents external to the company. Punctuality applied to programs is related to the time of connection and access, but it is also important in the stages of pre-production, production and post-production. The absence of mistakes is also considered a quality value, or failing that, the ability to rectify with immediacy, especially in the online context. To achieve this objective, forethought is essential, as it helps to anticipate and find solutions to any possible difficulty, as well as to be able to timely report to the people affected by any delay.

Technical quality also has to do with the development of abilities and skills to manage technical instruments (lights, cameras, sound mixers, etc.) and to perform well in front of the cameras for any of the television genres [JOH 05]. In the case of technical instruments, innovation may be an element of quality. Similarly, standards of good performance should be found in rhetoric and drama. In both fields, there are very specific guidelines that can be translated to television. A performance that is natural, true to the script and witty entails a professional skill that determines the quality of a program. The technical work must be at the service of intellectual work.

Along with technical aspects, we should highlight that marketing strategies in the economy of abundance are essential to publicize the launch of new products and reach the desired target.

Technical quality also depends on the bandwidth and the capacity of transmission. These issues are external to the content industry, but its success also depends on their efficacy and potency.

Furthermore, to entertain the audiences properly, a good story and a good script are essential; for knowing and being able to communicate reality, choosing or developing a credible, consistent and enriching script for the audience are crucial. Decisions on content demand respect for human dignity in its double dimension and respect for human nature and to individuals [SAN 00]. Respect should not only be present in the preparation of the content and the selection of topics and their approach, but must also address the public who sees the content and interacts with it. Sometimes, the viewers' intellectual and sensitive ability is underrated and this attitude ends up frustrating professionals who make the content. Those who do not respect others end up not respecting themselves.

Thus, studying the true interests of viewers becomes a necessary action for the pursuit of quality. For viewers to identify the content, programs must address what is close to their lives, their choices, their fun, their friends and professional and family relationships [SAL 10]. In this regard, it is worth investigating those issues that are of universal human interest. The human interest lies in issues that have an involvement in the recipient, and influence their decisions and affect them in some way.

Finnis [FIN 88] identifies seven values that constitute human goods sought after by people: life, knowledge, aesthetic experience, the game, internal harmony, harmony in relationships and the religious fact. These issues affect aspects close to the public, of universal and permanent interest. Other issues, more superficial and inconsequent, may attract the audience momentarily, but will hardly keep their attention with continuity over time. Furthermore, all these issues should be part of simple stories.

In everything that happens or arises, it is necessary to look for the relevant information and eliminate the superfluous data. In this sense, research about anthropological and social issues can be crucial [LAS 96]. New technologies promote dialog, which helps in finding out the tastes, interests, wants and needs of the public and the ways to meet them.

It also requires doses of wit and certain knowledge of what really interests people and makes them happy [GAR 00]. In recent TV festivals, the most popular television programs have been the *feel-good formats*. This shows that the viewers are looking forward to enjoying good stories that help them become better. Good stories that have travelled the world and have captivated audiences over time are simple stories, told in a simple fashion and appropriate to a target audience. Audience research has to be at the service of intuition and professional expertise. The professionals' training, their culture, common sense and ability to see can never be supplanted by other marketing techniques, no matter how stringent they are. Finally, the way to tell stories is more important than the stories themselves [PUJ 98]. For this reason, it is not intelligent to think that quality means exclusion of certain topics. On the contrary, we should not try to exclude certain issues but reflect on how to present them correctly and why.

These forms of storytelling must be adequate for the genre of the program. This is a key feature to ensure the communicability of the content. According to Valaskivi [VAL 00], the genre of a program is the result of a social and culture contract between the production team and the viewers. This contract is essential if the viewers are to recognize the implications that give meaning to the programs. A program can be said to have reached the expected quality when it is appropriate for its audience [BRO 92]. There is quality when the audience connects with the story narrated and feels involved with what happens, and this is only possible if the reading pact is respected, that is, if viewers know the references and the standpoint from which the issues are viewed [ENT 93].

Successful content aims to connect with the audiences' tastes and what they like [SAN 09]. In this sense, humor and sympathy are values that contribute to make more friendly and entertaining programs. Obviously, humor will be proportional to the genre of the program and the topic at question. However, if programs are not entertaining, it cannot be said that they are quality programs. Good humor does not hurt programs, nor awakens reticence or false grievances. In general, a smiling face is not bad for any TV show.

Furthermore, TV is a medium that exploits the external senses, sight and hearing, so the viewer is more sensitive to formal aspects that relate to the beauty and harmony. Crass and sensual things also impact sensitivity, but these factors eventually end up tiring the public and not entertaining. Therefore, the coarse and vulgar must be replaced by education, respect and pleasing to the senses. People either get used to the former and start asking for more intensity or just get tired of it, but it is harder to reject the latter.

Together with the aesthetic dimension, verisimilitude is an essential ingredient in any audiovisual content. Fiction is about the free actions of men and women and a real reference is needed for it to fulfill its function. If moral relativism were true, fiction would not work, because there would be no fixed rules to understand the reading pact. Communication professionals should aim to ensure the veracity of their work to reflect the truth without hiding an ideological approach, if there is one.

The visual nature of television has sometimes led to sublimate the image, apparent and sensitive above the intellectual, relevant and permanent [BOG 04]. The impact is more valued than the rational argument. Bourdieu [BOU 97] points out that journalists have the ability to turn the banal into extraordinary and Bayles [BAY 01] talks of "perverse modernism" whose presence is "deliberate", so that there is an attempt to make the ingredients of TV shows offend as much as possible. As in art, in the media, there is a desire to provoke, to always go a little beyond what is allowed in order to multiply sales.

The danger that comes with this approach lies in the separation of reality and the loss of credibility parameters that make media content understandable. The audience may lose sensitivity to put the facts into perspective and also the keys to understand the messages in tune with those who produce them. As Martín-Algarra [MAR 01] says, a frivolous approach to communication generates the opposite of communication: ignorance, isolation and disintegration of social ties.

Considering that the task is not easy, and that most cases require taking large risks and developing firm and highly innovative attitudes, an effective way to focus on quality is to find successful paradigms of quality in whatever context we live [HOL 03]. Those examples may inspire the creation of high-quality and highly successful content. Whoever promotes them will find in them the argumentative force to convince others of their convenience.

6.4. Bibliography

[ALB 96] ALBERS R., "Quality in TV from the perspective of the professional program maker", in ISHIKAWA,. S. (ed.) *Quality assessment of Television*, John Libbey, UK, 1996.

[ALB 12] ALBERS R., "Quality in TV from the perspective of the professional program maker", *Studies of Broadcasting*, no. 28, Mach, pp. 7–77, 2012.

[AZU 09] AZURMENDI A., *El Consejo Audiovisual de Navarra: sus funciones y competencias en el contexto europeo de regulación del sector audiovisual*, Consejo Audiovisual de Navarra, Pamplona, 2009.

[BAY 01] BAYLES M., "The perverse in the popular", *Wilson Quarterly*, Summer, pp. 40–47, 2001.

[BOG 04] BOGART L., "Reflections on content quality in newspapers", *Newspaper Research Journal*, vol. 25, no. 1, pp. 40–54, 2004.

[BOU 97] BOURDIEU P., *Sobre la televisión*, Anagrama, Barcelona, 1997.

[BRO 89] BROADCASTING RESEARCH UNIT, *Quality in Television. Programmes, Programme-makers, Systems*, John Libbey, London, 1989.

[BRO 92] BROWER S., "Fans as Tastemakers: viewers for quality television", in LEWIS L. (ed.), *The Adoring Audience: Fan Culture and Popular Media*, Routledge, London, 1992.

[COR 99] CORNER J., *Critical Ideas in Television Studies*, Clarendon Press, Oxford, 1999.

[COS 05] COSTERA I., "Impact or content? Ratings vs quality in public broadcasting", *European Journal of Communication*, vol. 20, no. 1, pp. 27–53, 2005.

[DIE 11] DIEGO P. *et al.*, "La percepción sobre la calidad de las series televisivas en España. Contraste entre el público y los profesionales", *Zer*, vol. 16, pp. 69–88, 2011.

[ENT 93] ENTMAN R., "Framing: toward a clarification of a fractured paradigm", *Journal of Communication*, vol. 43, no. 3, pp. 51–58, 1993.

[FIN 88] FINNIS J., *Natural Law and Natural Rights*, Clarendon Press, Oxford, 1988.

[GAR 00] GARCIA-NOBLEJAS J.J., *Comunicación borrosa*, Eunsa, Pamplona, 2000.

[GON 99] GONZÁLEZ A.M., *Expertos en sobrevivir*, Eunsa, Pamplona, 1999.

[GRE 72] GREEN T., *The Universal Eye, World Television in the Seventies*, The Bodley Head, London, 1972.

[GUT 00] GUTIÉRREZ C., "Televisión y calidad: Perspectivas de investigación y criterios de evaluación", *ZER*, vol. 5, no. 9, pp. 151–184, 2000.

[HIL 95] HILLVE P. "After Research, Measurement or Perhaps Vice Versa?", *NORDICOM Review*, no. 1, pp. 23–36, June, 1995.

[HOG 89] HOGGART R. (ed.), *Quality in TV: Programmes, Programme-makers, Systems*, John Libbey, London, 1989.

[HOL 03] HOLT J., "Vertical vision: deregulation, industrial economy and prime-time design", in JANCOVICH M., LYONS J. (eds), *Quality Popular Television: Cult TV, the Industry and Fans*, British Film Institute, London, 2003.

[HUD 87] HUDSON R., *Mass Media, A Chronological Encyclopedia of TV, Radio, Motion Pictures, Magazines, Newspapers, and Books in the U.S.*, Garland Publishing Inc., New York, 1987.

[ISH 96] ISHIKAWA S. (ed.), *Quality Assessment of Television*, University of Luton Press, Luton, 1996.

[LAS 96] LASAGNI C., RICHERI G., *Televisione e qualittá*, RAI-ERI, Roma, 1996.

[LEG 93] LEGGATT T., "Quality in TV: the views of professionals", in ISHIKAWA, S. (ed.) *Quality Assessment of Television*, John Libbey, UK, 1996.

[MAR 01] MARTIN-ALGARRA M., "Hablar para entenderse", *Nuestro Tiempo*, pp. 105–109, October 2001.

[MED 98] MEDINA M., *Valoración publicitaria de los programas de televisión*, Eunsa, Pamplona, 1998.

[MED 06] MEDINA M., *Calidad y contenidos audiovisuales*, Eunsa, Pamplona, 2006.

[OTS 08] OTS M., "Media and brands: new ground to explore", in OTS M. (ed.), *Media Brands and Branding*, Jönköping, Sweden, 2008.

[PEL 06] PELEEGRINI S., MUJICA M.C., "Valor Agregado Periodístico (VAP): la calidad periodística como un factor productivo en un entorno medial complejo", *Palabra-Clave*, vol. 9, no. 1, pp. 11–28, 2006.

[PIC 00] PICARD R. (ed.), *Measuring Media Content, Quality, and Diversity. Approaches and Issues in Content Research*, Suomen Akatemia and Turku School of Economics, Turku, 2000.

[PUJ 98] PUJOL C., *Tarea de escribir*, Pamiela, Pamplona, 1998.

[SAL 10] SALGADO A., "Una mirada a la creatividad televisiva", in *Creatividad en televisión: entretenimiento y ficción*, Fragua, Madrid, 2010.

[SÁN 00] SÁNCHEZ-TABERNERO A., *Dirección estratégica de empresas de comunicación*, Cátedra, Madrid, 2000.

[SÁN 09] SÁNCHEZ-TABERNERO A., *The Content of Media: Quality, Profit and Competition*, Media XXI, Lisbon, 2009.

[SHU 03] SHUTERMAN R., "Entertainment: a question for aesthetics", *British Journal of Aesthetics*, vol. 43, no. 3, pp. 289–307, 2003.

[TER 06] TERAMO M.T., "Calidad de la información periodística en Argentina. Estudio de diarios y noticieros", *Palabra-Clave*, vol. 9, no. 1, pp. 57–84, 2006.

[URR 08] URRETAVIZCAYA M., *La nueva televisión digital en el universo multimedia*, Universidad de Deusto, San Sebastián, 2008.

[VAL 00] VALASKIVI K., "Being a part of the family? Gender, genre and production in a Japanese TV drama", *Media, Culture and Society*, vol. 22, pp. 309–325, 2000.

[VIL 00] VILLANUEVA E., *Ética de la radio y de la televisión. Reglas para una calidad de vida mediática*, Universidad Iberoamericana, UNESCO, México, 2000.

[VON 10] VON RIMSCHA B., DE ACEVEDO M, SIEGERT G., "Securing quality in public service televisión entertainment", *Studies in Communication Sciences*, vol. 2, no. 10, pp. 7–25, 2010.

[WOL 99] WOLF M., *The Entertainment Economy. How Mega-media Forces Are Transforming Our Lives*, Penguin Books, London, 1999.

7

From Attention to Engagement in the Battle for the Audiences

In a media market characterized by fragmentation, it becomes increasingly difficult to define what success is. Ratings, shares, box office figures, sales or subscription numbers do not seem enough, as linking success only to large numbers of viewers seems unfair for niche markets and small groups of faithful long-time consumers. What ensures the continuity of a show or the contract for a new season? Like when discussing quality, it is difficult to point out exactly what is needed to gain the viewers' attention and keep it. The core of media business and the key to the audiences' heart lie in the content, however, many aspects of content are not always measureable. The ideal relationship to be established between an entertainment product and its public depends on the product and its nature and purpose. What seems clear is that in the fragmented ecosystem, a successful media product needs to be able to transform the attention gained from the public into a relationship of engagement that can also have a resonance in people's wallets.

In this chapter, we will review some of the elements that give way to the relationship of content viewers and suppliers in the media market, especially in the market of audiovisual entertainment.

Entertainment is a ubiquitous phenomenon, present in all cultures. It is defined, broadly, as an activity designed to bring diversion and fun and, to a lesser extent, instruct through stories [BRY 96]. One way to experience entertainment is through the media and audiovisual products, content of different genres in film, television and other media.

Chapter written by: Alicia URGELLÉS.

The popularity of the "audiovisual phenomenon" has been growing steadily since its inception. Although the business in its various forms may go through turbulences from time to time, there will always be people willing to produce and consume stories. It is even said that the need for stories grows when times of crisis arise, like when the popularity of the nickel movies in the United States grew during the crisis of 1929. Studies show how the state of the economy does not negatively impact the supply of and demand for cinema entertainment [VON 13]. Currently, according to the statistics of the Motion Picture Association of America, global box office for all films released in each country around the world reached $38.3 billion in 2015, an increase of 5% over 2014's total [MOT 16] (2015 Theatrical Market Statistics, Motion Picture Association of America).

With a special power to incorporate messages to life patterns of consumers and to form their perceptions and attitudes, the importance of entertainment has not gone unnoticed by the scientific community. From many disciplines and perspectives, academia has tried to closely follow the evolution of the phenomena surrounding the audiovisual entertainment. From the standpoint of media economy, audiovisual entertainment is of great interest due to the nature of the product and to the market relations established through it. Relations, as we discussed in previous chapters, in which time plays an important role, especially in the advertising-financed television, in which it works as a commodity.

Media, in general, work with people's time and attention. Limited attention has been a common issue studied in psychology and behavioral economics. Research over the attention economy emerged in the early nineties, around the problem of how people manage information in their daily lives: people allocate their capacity to process information in a given set of sources and their "agenda", which is also built from the issues and people who compete for their place in it. The approach to this topic in media economics is related to management and the usefulness of media today as content organizers. Society expects media to work as a way to manage, filter, edit, select content and help users to discern what is relevant.

Management of a scarce resource (attention management, in this case) requires decisions, which are influenced by the price and the value that is expected from the product. Abundance disempowers price as a central element in making consumption decisions, while shortage increases its economic value [PIC 10]. Despite analogies with conventional economic competition reflected in formulations such as "pay" attention, it is not possible for the provider of content to acquire full control of a person's attention, nor is it possible to control the exposure to the continues continuous calls for attention [FAL 07].

Attention has an instrumental value and a value in itself. As an instrument, attention is a means through which another end is achieved, for example, persuasion. Attention also has a value in itself; the value of the attention received and given that implies interest, appreciation, love, care, etc. According to Davenport and Beck [DAV 02], attention is a mental commitment.

7.1. Attention economy: the relationship with the creative mastermind

The notion of attention as a commitment leads us back to the idea of the relationship that takes place in the market. In law, a commitment is an agreement, a promise, a contract or pact in which parties agree to a series of actions and rewards. A contract exists when people agree on a common statement of intent designed to regulate their rights [CAS 08]. Such a contract occurs between the minds in need of information and entertainment and the creative minds seeking to meet that need. The said compromise affects the way both supply and demand will behave in the market.

The conceptualization of the entertainment market as a relationship goes beyond legal considerations and practices regarding copyright or fair competition issues. In the case of media in general, and audiovisual entertainment in particular, the network of relations that are created is relevant because they are what makes a dialogue possible [NIE 01].

In this case, the idea of the market as a relation is closer to the concept of the pact between a story's narrator and readers. As seen in previous chapters, the pact represents a coincidence of "codes" among the subjects. In the construction of a narrative, there are some codes that structure the story and, when they are common to both the narrator and the narratee, they allow a better understanding between the two, which results in a better message that can be shared more easily [SÁN 06].

In this relationship, consumers are as responsible as suppliers. It is certainly important that content producers know their target audiences and what they demand and need. However, viewers also need to be aware of and use those codes in the narrator–reader pact, intended to make believable what is known to be true about human actions. It is part of everyone's audiovisual literacy to minimally know the resources of poetics and recognize that they are capable of setting a public record of actions that are socially and culturally memorable, through the configuration of possible worlds and universes [GAR 05].

As for the producers and suppliers of content, it is good to highlight to the special status of "professionals" they have. Status establishes a dynamic in which one of the subjects has a role as an authority over the other [TOR 09].

Postman [POS 84], from his criticism and skepticism of television as a place for relevant conversations, invites us to think about the question of authority in the media. "To talk seriously about television, one must therefore talk of epistemology" [POS 84, p. 17]. Certainly, in a relationship in which the product that binds the parties has such an implication in the way of understanding reality – a relationship of knowledge – it is important to consider some inherent issues, like concepts of "authority" and "trust": the authority of those who produce content, and the trust that consumers have in that authority.

The aforementioned relationship of trust is necessary for the proper functioning of the relationship [ARR 04], even from the point of quality of the entertainment experience. If there is confidence in the result of the work of professionals, the audience will experience better immersion in the proposed universe.

Today there is much talk in marketing and management about loyalty and brands trying to inspire loyalty in their customers. Loyalty, however, is something customers can also demand from brands. Media brands must be committed to their audiences and offer a determined service with a determined quality. Media brands specialized in offering audiovisual content should determine the standards they want to maintain and be loyal to themselves and their work as a way to be loyal to their customers.

As explained by Torregrosa and Gutierrez [TOR 09], every profession originates from a learning process and gains recognition, which generates expectations about the behavior of their professionals related to an ethic of service. A profession stands on the idea of competence and authority (technical and moral) regarding an activity or sector, and the professional is therefore someone who is in position to occupy a consecrated place in society [BEL 76].

The production of an entertainment product implies an act of prudence carried out by the media professionals, who become committed to the true nature of their actions and directly involved in the kind of discourse that they choose in order to communicate those actions in the form of a poetic object: the media product [TOR 09]. Entertainment professionals combine several dimensions of practical activity in the production of their contents: rhetoric, poetics, aesthetics, ethics and politics. As Garcia-Noblejas explains, a "poetic truth" is the result of the decisions that were made on what is known as true [GAR 05].

In the activity of screenwriting, for example, Brenes explains how the work of writers is "not only a technical matter focusing on the suitable use of narrative tools, (...) it is a human activity in which other dimensions come into play. These dimensions have to do with the authors' and spectators' conscience and with their status and dignity as human beings that write or see screenplays" [BRE 11, p. 104].

Good professionals will properly combine the practical dimensions in the product so that, when received by the audience, the creation can produce knowledge that is also wisdom. Narrative constructions create a universe of discourse in which the viewer must accept the authority of the narrator, who has established the meaning of the proposed universe, its characters, actions and circumstances. The acceptance of the constructed universe is not a submission; it is part of the pact between the reader and narrator. Unlike other ways of communicating, these creations are not intended to show the truth directly. They seek to be "poetically plausible" [GAR 05, p. 259] so that whoever consumes it can find themselves in the characters and their actions. That is, "stories are about the likeliness of actual worlds. Even though stories do not offer verifiable truth, people tend to perceive reality from fictional events, actions and relationships in stories" [AHN 12, p. 10].

Identification with the characters of fiction frames and the desire to live many different lives by adopting the positions of the different characters are central components for enjoyment, impact and appeal of audiovisual productions. "A story with a high degree of perceived reality leads the viewer to be involved in the story, and psychological involvement with the story, in turn, functions as a predictor of enjoyment" [AHN 12, p. 11]. For the study of the entertainment industry, the knowledge of the reasons that stimulate the audience to consume gives clues to advance on the effectiveness of communication relationship and the efficiency of the sector.

That is why understanding of narrative and dramatic works is necessary to understand the economics of entertainment. Managers of media businesses cannot think it is enough to know about management, because the specificity of the entertainment world makes it essential for everyone involved in it to understand the mechanics and affairs of storytelling. The message is the core of the connection between the agents of any communication process, and a significant message is the key to a significant connection.

Stories have great communicative power. We live surrounded by stories, in films, television, advertising, literature and media. They are benchmarks in which we recognize our own problems and states of mind when we see them reflected in someone else. This identification is what really attracts and keeps the attention of the viewer.

As Medina [MED 98] argues, consumption is not a business relationship, but its consequence. The purchase or consumption is the acceptance of the offer by the demand. Herrero [HER 07] also points out that, by its very nature, the business relationship involves reciprocal benefits, and ultimately pursues that the relationship ends in consumption.

The content suppliers' search for a significant relationship with their viewers is embodied in the pursuit to meet the needs of the audience. Satisfaction is central to the economic viability of a media company, because satisfied customers are more likely to remain loyal to the brand, while dissatisfied ones will flee to the alternatives [PAT 11]. The effectiveness of the work of an audiovisual service will find its completion, its accomplishment, in a relationship of engagement.

7.2. Engagement as the completion of the relationship

The term "engagement" is used in marketing, which refers to a significant bond between a brand and its stakeholders, who are to feel some emotional involvement or commitment toward the brand's identity and, consequently, toward its products or services. Such a relationship of engagement is especially important for brands dealing with experience goods and credence goods like media products, which tend to behave unpredictably in the market. Engagement has been viewed as a concept that may contribute to a better understanding of service performance and customer loyalty outcomes [BOW 09, GAM 10, HOL 11].

Through the pursuit of engagement, the brand puts a special emphasis on building a relationship of dialogue with its stakeholders, a conversation that brings the brand closer to its public and manages to not only draw their attention but also keep it. "Companies are now realizing that engagement is also a more strategic way of looking at customer and stakeholder relationships. In this emerging approach, engagement refers to the creation of a deeper, more meaningful connection between the company and the customer, and one that endures over time. Engagement is also seen as a way to create customer interaction and participation" [KUM 10, p. 297]. Although the term is widely used to refer to an increase in customers' spending, to the media it is more than that. The goal of engagement must be in the design of the business or it will just serve as makeup, or as a sporadic strategy to increase sales.

Although it is not possible to explain exactly what leads people to choose where to invest their time and attention, Davenport and Beck [DAV 02] suggest that to gain and keep the attention of people, it is necessary to involve emotions. Consciously or unconsciously, emotions are references that relate everyone within the world. Moreover, from an economic perspective, emotions have two virtues. First, they arouse the attention and second, and at a deeper level, they help build a closer relationship.

Engagement can be reached through interaction and participation. In the context of the use of narrative media, and following the theory of narrative transportation, participation can be conceptualized as the "absorption" in a narrative [AHN 12]. This absorption would be the vicarious experience of events and emotions of a fictional story as if they were real. According to this theory, the person transported to the story goes through a focus of attention in this story and the reality it describes. This concentration on the world of the story leads to distancing from the immediate reality, which is a means of distraction to escape stress and worries.

Furthermore, participation begins with understanding the story. Understanding is the step that allows the viewers to put together the story with their own experience in order to make it relevant and give strength to the bond between the consumer and the content. Understanding the history activates the immersive experience, unless a feeling of incongruity interferes with immersion. Such a feeling of unreality will happen if there are failures of internal coherence or external realism in the content, that is, if there are failures that affect the overall quality of the product. Moreover, as explained by Grandío [GRA 09], a universe that is pleasant to the viewers can make the viewing easier.

Participation goes further than immersion with the possibility of user intervention. Jenkins [JEN 08] talks about the game mentality that exists in fans' communities and the organizational skills present in many users and viewers, as well as the role they can play in many processes that had been only for media professionals. This democratization is presented as a break from the traditional relationship between the broadcaster and the spectator. Consumers can use "new media" to get involved with the content and move past the monologue of "old media".

In the English language, which has created some neologisms to describe the new modes of consumption and phenomena of the media market, the word producer is used to identify the user who launches video production; a prosumer is the consumer who contributes to promoting and improving the brand image by participating in forums, social networks and sharing their consumer experiences. A viewser is the user who comments on videos and other contents [TAP 08]. What these neologisms show is that interaction with content and participation in media production are becoming more common among users of online media. This causes the monologue of old media to give way to a content market in which producers and consumers can gather and be closer than before.

This happens thanks to the convergence in the media industry. Convergence, according to the Merriam-Webster dictionary, refers to the merging of distinct technologies, industries or devices into a unified whole. In the case of media, the process of convergence can be studied from many perspectives and levels; on an industrial level, organization level or activities level, but overall, it implies a change in both media production and media consumption modes.

The entertainment industry has embraced the idea of convergence, and one example of it relates to the content. These are the brand extensions, multimedia and synergies as well as franchises that fill the market cross-media they allow the content to be enjoyed in a wider spectrum of spaces, devices, moments and media. The key of them all is always the content.

Suffice as an example what is currently happening in the television industry in the United States: according to Jenkins, Ford and Green, a successful series attracts almost the same number of illegal downloads as of television viewers [JEN 13]. These authors find in this phenomenon the passage from a paradigm based on television schedules and appointments to one based on engagement: the audience become a group of agents whose job is to generate alternative forms of market value [JEN 13, p. 115]. The "committed" audiences are measured differently from the traditional ones because they act differently. For these consumers, the content is more important than the platform or consumer context. For example, content-related factors are the top drivers of Netflix and Hulu Plus subscriptions: "over 70% of service subscribers state that they have these services due to specific titles available through the service, and over one-third of consumers do so in order to access original content" [PAR 16, p. 5].

It is with these audiences with whom it is possible to engage in a "transmedia engagement" based on the content that can expand ("spreadable" content) and go beyond the first and shallow consumer relationship [JEN 13]. According to Jenkins [JEN 08], we are witnessing the emergence of new narrative structures: synergetic and transmedia, the entertainment for convergence and for the era of collective intelligence.

In his work, *Convergence Culture*, Jenkins [JEN 08] puts special emphasis on the issue of collective intelligence and participatory culture. Jenkins speaks of media consumption, mainly entertainment, as a collective process. It shows how viewers leverage individual knowledge in virtual communities to create a collective intelligence. They create theories about their favorite series and interact with the content they receive in a much more active form. The relationship with the universe of the series changes to have a challenge dynamic. A well-achieved dynamic of challenge and game will always facilitate enjoyment.

In other words, it could be said that engagement is a result of enjoyment: insofar as content is enjoyed (in the sense we have previously seen of immersion, participation, understanding, etc.), a significant relationship between the subjects of the market will be established. The idea of a relationship of engagement summarizes the purposes of the audiovisual entertainment as an experience product, creative good that satisfies the need to invest well our free time.

Doyle [DOY 10] explains how, in the UK television industry, they have a way of working called "360-degree commissioning", which is a strategy regarding the concept that implies "that, from the earliest stages of conceptualization, content decisions are shaped by the potential to generate consumer value and returns through multiple forms of expression of that content and via a number of distributive outlets (e.g. online, mobile, interactive games and so on) of which conventional television is just one, albeit still a very important one" [DOY 10, p. 432].

What can be taken from this is the significance of caring for the contents that constitute entertainment products, the development of the conflict and the characters that comprise them. This would also lead us toward the debate about what quality is discussed in previous chapters and its relationship to the tangible (and intangible) success of an audiovisual product, like a TV program or film.

In the engagement situation, the relationship between the content creator and the person who enjoys it is thus successfully completed; a relationship that has needed the time and attention of the consumer, as well as the authority of the professionals involved and confidence of the audience among the terms of that relationship.

The theory behind what needs to be done to keep consumers engaged with a media brand has practical implications beyond the quality of the story and its capacity to immerse viewers in it. While the core of the business in the world of audiovisual media will always be the content, the product, it is also necessary to consider the quality of the service provided and the extra features related to that content that can be offered for a richer and more engaging consumption experience.

7.3. Trends and practical implications

It is often said that the content is more relevant than the channel or platform that gives access to it, because as soon as the content changes "continent", its viewers and fans will move with it. The talent of the creative team behind the audiovisual product leaves a recognizable mark in their work that attracts viewers and keeps them engaged. It is important for any company involved in audiovisual content to recognize that value and the relevance of authorship.

The viewers' loyalty tends to remain with the programs, presenters and artists, over the network that hosts them, which is why these companies must take good care (legal, economic, human) of the creative talents they manage. Authorship can give way to powerful brands with a strong production value to call the viewers' attention. It also constitutes the know-how of the business and the spine of the industry's memory. The "memory" of a company is its most precious intangible asset, and it is the sum of the intelligence of its employees over time [NIE 00].

What the company gets in return for the investments on creativity and talent include economic benefits and the know-how that has been learned over projects. Know-how and memory push the industry forward, by pushing quality forward and growing in experience and inspiration for a new project. Memory is also important to create a sense of community, of collective memory: a "popular culture" among the professionals in business and the viewers that recognize their authorship and authority. Media businesses can look in their quantitative success for the mark they are leaving in the collective identity of the industry.

This well-managed creative talent will help design new forms of exploitation of content. We are living in a very exciting time, in which possibilities are infinite and the world of content, thanks in part to the advances in technology, is very open to experimentation. A simple reversioning or reuse of media products is not enough for this approach to content. Viewers nowadays are "highly-engaged multi-platform consumers who crave specific content that appeals to their interests, hobbies or passions throughout the day, no matter the time or device" [PAR 15]. This approach asks for a continuous management of content and a continuous relation with consumers, throughout devices and times of the day, in a way that the content can accompany and serve the consumer at their will. This constitutes a great opportunity for lifestyle or interest-based content around categories like cooking, finance, sports, fashion, fitness and more.

The media world right now can learn a lot from other industries of goods and services. Digital economy has blurred the boundaries between industries and business ideas and, when in search of inspiration for innovation, media need to broaden their interests and be open to suggestions from even unexpected sectors; from the worlds of online retailing, software development, hardware, tourism, restoration, transportation or banking. While the content is essential for the relationship with the audience, it is good to remember that media is more than their products; they are a service and the management of the said service must ensure a satisfactory experience for the user. Some ideas needed to enrich that experience might be found in other sectors.

Something that is normal nowadays is the fact that media travels with users in their pockets and is completely mobile; the mobile consumer has almost no space limitations and a lot can be done regarding that mobility. The quality of the mobile experience needs to be taken into account, and also all the opportunities that can arise from being connected to users 24/7 and in the palm of their hands.

Users are localized all the time, thanks to their mobile devices. Even though this fact may awake some paranoid instincts in us, it is still a fact and geolocalization is a tool that will be more and more exploited. This will turn into the ubiquity of media, which can translate into interesting projects, like the "Pokemon Go" game released in summer 2016. This game works like a free-to-play app in which players use their mobile devices to catch and play with virtual creatures they have been able to find during a location-based augmented reality endeavor. The official website explains how "as you walk through the real world, your smartphone will vibrate to let you know you are near a Pokémon" (Pokémon Go website). The game became a fast success, with its associated controversy, problems and issues. Whether it proves to be a good idea or a failed experiment in the long term, this game is just another example of what can be done to engage with consumers. And as a proof of this engagement being good for business, news point out a rise in market value of the firms responsible for the game along with the game having been the most downloaded free app in the mobile apps store [KIM 16].

The game dimension of entertainment media also offers an interesting area for innovation and an opportunity for engagement. The spirit of play is essential in the development of a culture, since it works as a support for social aspects such as politics, debate, social relations, military strategy or the arts; through games, we learn how to make order or understand the economy. The game is also closely connected to narrative, as it is free time devoted to a social representation on a smaller scale; a training for life. As Squire [SQU 06] notes referring to video games, a game is a designed experience: participants foster new identities within the "universe" of the game, "players' understandings are developed through cycles of performance within the gameworlds, which instantiate particular theories of the world (ideological worlds)" [SQU 06, p. 19].

Immersive interactive digital entertainment has become an important means of exerting an enormous economic, cultural and social influence. Squire explains that gaming experiences influence attitude toward work and learning, so its current manifestations (for example, in audiovisual entertainment, video games) represent a shift toward a "culture of simulation" that makes it possible to construct and participate in hypothetical worlds [SQU 06].

In line with this gaming dimension, media can use the tools technologic advances provide to expand stories and characters and bring them closer to their consumers. Virtual reality and augmented reality are just examples of tools that can be exploited in benefit of contents while working as a commercial success. The newer wave comes in the form of the so-called "Internet of Things" (IoT), which is still in early stages of development, but offers much potential for entertainment media: Internet connections embedded into any device, from cars, kitchen, security systems, televisions or anything, can be used to communicate, collect and exchange data. This offers a higher level of personalization and many more points of connection between providers and consumers of content.

Actions that rely on a brand's good solid content tend to have good influence over its market value. Brand extension, merchandising, sequels and prequels, crossovers and remakes are common terms in the audiovisual industry nowadays. Being able to take advantage of a fictional universe that is interesting and attractive for a consolidated fan base is an undisputable resource. Again, creative talent, along with a good market research, are needed to exploit the universe correctly and avoid angering fans or causing franchise fatigue.

A media brand's characters, worlds and stories are very important assets and a part of its "memory" and know-how. It is not only a chance to interact with viewers that are already engaged, but also an opportunity to attract new consumers. For example, when the last movie of the "Star Wars" saga was released in December 2015, it brought together fans that had begun gathering in 1977 when the first movie came out, which gave Disney "its best quarter in company history" [GRE 16].

Sequels, prequels and remakes seem to be a big trend in today's worn-out film panorama. But this trend uses the feeling of familiarity of a good story with well-known characters, as well as the viewers' nostalgia and the filmmakers' wish of revisiting or retelling stories to add their own twist. That element of surprise and expectation seems effective to attract engaged fans and curious new consumers. Moreover, there is a sense of community that accompanies sagas or television series. It keeps viewers excited about new episodes and drives them to watch as a part of a ritual that allows them to share their comments with other members of the community.

Another dimension of audiovisual products that has changed over the years is time. Video on demand has changed the viewers' perspective over times of viewing and has also changed their perspective over lengths of viewing.

Phalen and Ducey [PHA 12] run through various options of measurements for engagement related to lengths of viewing, proposed by several scholars: the percentage of minutes viewed, the proportion of the audience that watched 80% or more of a program, the time spent with an individual channel, etc. They consider some of these measurements problematic, as the time spent in front of any screen does not necessarily indicate attention to what is being viewed. However, the length of viewing can still be considered a common representation for engagement: "the more time a viewer spends with content, the more "engaged" they are by it" [PHA 12, p. 141].

With traditional television, the length of viewing was determined in a great part by scheduling and programming. Many TV channels' strategies aim to retain the viewers' attention from one program to the next, or from one episode of a series to the next episode the next week. With video-on-demand, viewers are not limited by the channels' schedules and the phenomenon of so-called "binge-watching" has grown.

This tendency to binge watch is also interesting. It is not a new trend but it has turned to be easier to do with online viewing. The use of the word "binge" carries a slight negative connotation of "too much", which raises many questions about responsible viewing nowadays. The amount of time viewers spend watching an audiovisual content, especially television, has always been a cause of social concern, with more interest in the cases of young people and children. For television executives and advertisers, an increase in the time spent in front of the television screen – and others – is good news, particularly in young audiences. In this matter, responsible lengths of viewing also have a relation with the type of content watched, as not all products require the same attention or time, and "binge-viewing" is not meant for all contents.

"Not all series are enjoyed the same way" [NET 16]. The streaming service Netflix developed what they have called "The Binge Scale", after having analyzed more than 100 serialized TV series across more than 190 countries between October 2015 and May 2016:

> "The research examined member completion of the first season for all series. Data was only included for accounts that fully completed the season. Completion rates were organized into days and hours. The global median days to complete the first season of these series was five days. The median hours per session for completers overall was two hours and ten minutes. Series viewed less than two hours per day were identified as 'savored'. Series viewed more than two hours per day were identified as 'devoured'" [NET 16].

What is interesting about this research is the understanding of the narratives that lie underneath each type of content, and how that narrative influences the viewing experience and the response of the viewer in terms of binging or not: "it's no surprise that complex narratives, like that of *House of Cards* and *Bloodline*, are indulged at an unhurried pace" [NET 16].

Netflix' knowledge of their consumers' streaming behavior is a paradigm of the use of data to commission content tailored to users' taste. In a way, it is another form of participation of viewers, as they know their consumption habits are being monitored as a way to supposedly better tend to their needs. Users can rate and recommend contents and their decision to view determined products or not can help in the development of new products or existing ones.

Viewers' input is always important and more so nowadays with the fragmentation of the market. The process of testing pilots, for example, and the subsequent process of deciding whether programs live or die are complicated and most shows do not test well: some may not have mass-market appeal and good results in traditional testing, but still be able to draw smaller loyal audiences in niche channels [FLI 16].

Another study by Netflix explored at what point on a show people become "hooked" with it. The study determined that it was not on pilots; a pilot might get viewers interested, but that it takes more episodes for viewers to find a point in the story that compels them to come back and finish the season [SPA 15].

Storytelling is seen in marketing as a strategy to connect emotionally with consumers and bring them closer to a product with a story that can put the said product or brand and its role in perspective for the potential user. The opportunity to create value with stories is something many businesses pursue. The world of audiovisual entertainment has an advantage, as contents are the core of its business and investing in improving those contents is not a bonus or a side project. Money invested in market research and in development of better and innovative ways to connect with the audiences is money well spent, even if short-term results do not show that. The media industry must learn from these branding strategies related to storytelling that an important connection could be set between the elements of a story and the brand identity and the consumers' identity. Therefore, the attentiveness to the stories a media brand shares is the attentiveness to its brand identity.

At the beginning of the chapter we questioned what success meant in the audiovisual entertainment industry. Amid the current fragmentation of the offer, content creators and distributors must not forget about building a long-term relationship with their audiences. Sometimes, short-term hits may be useful to grab people's attention momentarily but, if not in line with the brand's identity and quality, hits may be unfavorable in inspiring confidence and authority in the audiences. It is also important to remember that, like in relationships between individuals, trust takes time to build but can be lost and broken quickly, and the audiences' expectations about the good authority of a brand rise with its good work. Once a media brand has found a good positioning with its audience, the quality of its work must remain high in order to maintain the identity, mission and vision it has committed to offer.

It seems that only big consolidated studios or networks are capable of turning engaging ideas into market triumphs. However, it is important to remember that smaller productions with smaller budgets but powerful stories are sometimes even more engaging (and profitable) than highly marketed series.

We have reviewed some of the elements that make possible the relationship between supply and demand in media, that is, between the content creator and the person who enjoys those contents. The market is where suppliers and purchasers of goods or service interact, their meeting point. A meeting means a connection, and in the case of media, the market represents a relationship "between minds in need of information and entertainment and creative minds seeking to meet that need" [NIE 01, p. 117].

The elements of this very particular market invited to propose some final considerations about the success and quality in this market. Garcia-Noblejas explains that the artist is responsible for their work to promote social and cultural fecundity, and not simply an effective tool in only ideological or economic enterprises [GAR 05]. In the claim or mandate of this approach, it is possible to observe the widespread confrontation between culture and economy, between what is considered success or effective from the cultural perspective and from the economic one, as if the "effectiveness" of the economy does not contemplate, by any means, an "effectiveness" in socio-cultural terms. In short, this is a chapter on the importance of contents in creating valuable relationships with the public. It is about the importance of entertaining and engaging content in the battle for the audiences.

Postman's fear about television not being a medium serious enough for good conversation is true around the content that gets good audience figures despite the lack of quality. But in these cases, the use of the word "success" is not entirely accurate. Success is relative to achieving a goal, the ability to work or attain a particular result. The true purpose and sense of media are directed toward knowledge: the need for enjoyment and comprehension of reality that knits inside the viewers' intelligence with their search for identity and fulfillment. If the objective pursued with the product is not according to proper satisfaction of this need, effectiveness and success are not complete. What the socio-cultural arguments consider a positive externality will also be one for economic standings, if the framework used to consider the success is more than just commercial.

According to Sanchez-Tabernero [SÁN 08], "any business success story is based on listening, understanding and responding to public demands" [SÁN 08, p. 266]. In addition, addressing the needs of people means to always keep the focus of the industry in its sense and purpose. The desire to fill free time with value, so that the consumption of entertainment becomes an investment more than an expense, is what will motivate the industry to maintain a high level of competition. As Nieto says, "spending becomes an investment when time leaves its mark on the life of intelligence, enriches the person by incorporating ideas, judgments, opinions, ideas, entertainment, with a profitable present and future" [NIE 01, p. 138].

7.4. Bibliography

[AHN 12] AHN D.H., JIN S.A.A., RITTERFELD U., "'Sad movies don't always make me cry'. The cognitive and affective processes underpinning enjoyment of tragedy", *Journal of Media Psychology: Theories, Methods, and Applications*, vol. 24, no. 1, pp. 9–18, 2012.

[ARR 04] ARRESE Á., "Algunas consideraciones sobre la gestión de productos y contenidos de los medios", *Comunicación y sociedad*, vol. 17, no. 2, pp. 9–44, 2004.

[BEL 76] BELL D., *El advenimiento de la sociedad postindustrial*, Alianza, Madrid, 1976.

[BOW 09] BOWDEN J.L., "The process of customer engagement: a conceptual framework", *Journal of Marketing Theory and Practice*, vol. 17, no. 1, pp. 63–74, 2009.

[BRE 11] BRENES C.S., "The practical value of theory: teaching Aristotle's poetics to screenwriters", *Comunicación y sociedad*, vol. 24, no. 1, pp. 101–117, 2011.

[BRY 96] BRYANT J., ZILLMANN D., "El entretenimiento como efecto de los media", in *Los efectos de los medios de comunicación: investigaciones y teorías*, Paidós Comunicación, Barcelona, 1996.

[CAS 08] CASADO L., *Diccionario de Derecho*, Valletta Ediciones, Buenos Aires, 2008.

[DAV 02] DAVENPORT T.H., BECK J.C., *The Attention Economy. Undestanding the New Currency of Business*, Harvard Business Press, 2002.

[DOY 10] DOYLE G., "From television to multi-platform. Less from more or more for less?", *Convergence: The International Journal of Research into New Media Technologies*, vol. 16, no. 4, pp. 431–449, 2010.

[FAL 07] FALKINGER J., "Attention economies", *Journal of Economic Theory*, vol. 133, no. 1, pp. 266–294, 2007.

[FLI 16] FLINT J., "Test audiences can make or break new TV series", *Wall Street Journal*, available at: http://www.wsj.com/articles/test-audiences-can-make-or-break-new-tv-series-1463244543, 15 May 2016.

[GAM 10] GAMBETTI R.C., GRAFFIGNA G., "The concept of engagement: a systematic analysis of the ongoing marketing debate", *International Journal of Marketing Research*, vol. 52, no. 6, pp. 801–826, 2010.

[GAR 05] GARCÍA-NOBLEJAS J.J., *Comunicación y mundos posibles*, Eunsa, Pamplona, 2005.

[GRA 09] GRANDÍO M., "El entretenimiento televisivo. Un estudio de audiencia desde la noción de gusto", *Comunicación y sociedad*, vol. 22, no. 2, pp. 139–158, 2009.

[GRE 16] GREENBERG J., "Star Wars gives Disney its best quarter in company history", *WIRED*, available at: http://www.wired.com/2016/02/star-wars-gives-disney-its-best-quarter-in-company-history/, 9 February 2016.

[HER 07] HERRERO SUBÍAS M., *Mercado de la televisión de pago en España. Canal Plus (1990–2000)*, Ulzama ediciones, Pamplona, 2007.

[HOL 11] HOLLEBEEK L.D., "Demystifying customer brand engagement: exploring the loyalty nexus", *Journal of Marketing Management*, vol. 27, nos 7–8, pp. 785–807, 2011.

[JEN 08] JENKINS H., *Convergence Culture. La cultura de la convergencia de los medios de comunicación*, Paidós, Barcelona, 2008.

[JEN 13] JENKINS H., FORD S., GREEN J., *Spreadable Media*, New York University Press, New York, 2013.

[KIM 16] KIM C., YAMAZAKI M., "Nintendo shares soar as new Pokemon mobile game captures hearts", *Reuters*, available at: http://www.reuters.com/article/us-nintendo-stocks-idUSKCN0ZO059, 8 July 2016.

[KUM 10] KUMAR V., AKSOY L., DONKERS B. *et al.*, "Undervalued or overvalued customers: capturing total customer engagement value", *Journal of Service Research*, vol. 13, no. 3, pp. 297–310, 2010.

[MED 98] MEDINA M., *Valoración publicitaria de los programas de television*, Eunsa, Pamplona, 1998.

[MOT 16] MOTION PICTURE ASSOCIATION OF AMERICA, *2015 Theatrical Market Statistics*, 2016.

[NET 16] NETFLIX MEDIA CENTER, "Netflix & Binge: new Binge Scale reveals TV series we devour and those we savor", Press Release, available at: https://media.netflix.com/en/press-releases/netflix-binge-new-binge-scale-reveals-tv-series-we-devour-and-those-we-savor-1, 8 June 2016.

[NIE 00] NIETO A., *Persona, Relación, Marketing*, School of Information Sciences, Austral University, Buenos Aires, 2000.

[NIE 01] NIETO A., "Economía de la apariencia y mercado de la información", *Comunicación y Sociedad*, vol. 14, no. 2, pp. 117–142, 2001.

[PAR 15] PARKE J., "The rise of the super viewer and why advertisers should care", *comScore blog*, available at: https://www.comscore.com/lat/Insights/Blog/The-Rise-of-the-Super-Viewer-and-Why-Advertisers-Should-Care, 29 September 2015.

[PAR 16] PARKS ASSOCIATES, *The OTT Playbook 2016*, 2016.

[PAT 11] PATWARDHAN P., YANG J., PATWARDHAN H., "Understanding media satisfaction. Development and validation of an affect-based scale", *Atlantic Journal of Communication*, vol. 19, pp. 169–188, 2011.

[PHA 12] PHALEN P., DUCEY R.V., "Audience behavior in the multi-screen video-verse", *The International Journal on Media Management*, vol. 14, no. 2, pp. 141–156, 2012.

[PIC 10] PICARD R.G., *Value Creation and the Future of News Organizations. Why and how Journalism must Change to Remain Relevant in the Twenty-First Century*, Media XXI, Lisbon, 2010.

[POS 84] POSTMAN N., *Amusing Ourselves to Death. Public Discourse in the Age of Show Business*, Penguin Books, New York, 1984.

[SÁN 06] SÁNCHEZ NAVARRO J., *Narrativa Audiovisual*, Editorial UOC, Barcelona, 2006.

[SÁN 08] SÁNCHEZ-TABERNERO A., *Los contenidos de los medios de comunicación. Calidad, rentabilidad y competencia*, Deusto, Barcelona, 2008.

[SPA 15] SPANGLER T., "Netflix data reveals exactly when TV shows hook viewers – and it's not the pilot", *Variety*, available at: http://variety.com/2015/digital/news/netflix-tv-show-data-viewer-episode-study-1201600746/, 23 September 2015.

[SQU 06] SQUIRE K., "From content to context. videogames as designed experience", *Educational Researcher*, vol. 35, no. 8, pp. 19–29, 2006.

[TAP 08] TAPSCOTT D., *Wikinomics: How Mass Collaboration Changes Everything*, Portfolio, New York, 2008.

[TOR 09] TORREGROSA M., GUTIÉRREZ R., "Saber y poder en los profesionales de la comunicación. Una aproximación al concepto de autoridad desde la epistemología de la comunicación", *Palabra Clave*, vol. 12, no. 2, pp. 287–299, 2009.

[VON 13] VON RIMSCHA M.B., "It's not the economy, stupid! External effects on the supply and demand of cinema entertainment", *Journal of Cultural Economics*, vol. 37, no. 4, pp. 433–455, 2013.

8

Monetizing in the Digital Era

The new digital economy has broken the barriers of high budgets, among other things, because it is still an economy of gratuitousness, where the advertiser is the one who pays. The ease of access and use of the Internet lowers prices. Micropayment has become a source of income when millions of users are willing to pay to see the end of a series, or an episode whose ending they have helped design.

Moreover, television has become a medium to pay for, and in which the viewer has emerged as ruler of his consumption decisions. New technologies allow the user to become an active subject of the right to information. This suggests a new freedom of expression, which is not only exercised by professionals, or guaranteed by private entrepreneurship, but also implemented by the consumer who can disseminate information, or the subscriber who, through his or her subscription, can express judgment on the quality and service offered. It is now more difficult to meet quality standards because the viewer has become a more critical judge when evaluating the quality of the content received or the content he or she can produce.

Chapter written by Mónica HERRERO.

8.1. Traditional funding systems of broadcasting

The three main funding systems for the broadcasting industry have been public funding, advertising and direct viewer payments. The most important in terms of revenues is advertising. Its goal is maximizing audiences. That is why Napoli [NAP 10] talks about "audience products". The media products can be described as dual goods [PIC 89] with two basic dimensions: it is content for audiences and generation of attention-time of these audiences for advertisers, so the audience enjoys free-of-charge television programming by giving part of its time [NIE 00].

To facilitate the buying and selling of advertising, the media market needs accurate information about the audience [POR 07] [WEB 14]. This data sets prices so it serves as currency for the advertising market [BUZ 12]. For TV, the predominant system of measuring the audience is based on meter panels, as those managed by *Nielsen*, *Kantar Media*, *Gfk* and *Ipsos* [POR 15].

Moreover, direct payment for television products (subscription, pay per view and any other form of video on demand) introduced price discrimination into an industry previously considered to be of free and universal access. As an immediate consequence, the income that is received does not depend on audience rating but rather the number of subscribers. Although this figure is subject to daily fluctuations, the channel has a contract-based income that allows it some room to maneuver. The advertising-supported channel, on the other hand, bases its contract with the advertisers on the viewer rating, with the drawbacks that have been described.

This requires going back to the very nature of audiovisual products as public goods. A public good is independent of the number of people who consume it, that is to say, the available amount of the commodity is not reduced by an increase in the number of consumers [PIC 89]. From this, we can deduce that consumers find themselves in a non-rivalry situation in terms of their enjoyment of the same commodity [HOS 97].

Given the nature of audiovisual products as public goods, a series of consequences arise that condition managerial decisions in the television industry. The high costs and the almost null marginal cost are the reasons why scale economies operate strongly in the audiovisual industry [PIC 89]. Program producers try to distribute their products among the largest possible audience, and therefore, the cost-per-viewer can be reduced. This results, for example, in distribution strategies or windowing, which intend to achieve the maximum economic profit of a single product by distributing it in different windows and at different times [OWE 92]. These repetition strategies prove that the value of a television product does not come to an end with its first broadcasting [DUN 90].

As Frith assesses, "by and large radio/television culture is (like the Internet, but unlike publishing) based on the expectation that programmes are generally accessible and not subject to price restriction" [FRI 00]. Furthermore, the nature of television programming as a public good hinders effective price discrimination, due to the uncertainty of the demand and the virtually nonexistent marginal costs. The television industry could be said to be in a situation where it lacks cost competition, as occurs in other cultural sectors. Until the advent of pay television, the audience had never had to consider cost when it came to choosing a particular channel or program and enjoyed a certain free-of-charge feeling as far as television programming was concerned [NIE 84].

Most of these characteristics of the audiovisual products are taken to the full in the case of their diffusion through the Internet. Their nature as public goods is reinforced, given the flexibility and universality of the Internet as a means of distribution, as well as the absence of geographical and legal barriers. The marginal cost is null, making it possible to develop economies of scale and windowing strategies in a more efficient way than ever before. Furthermore, apart from free access to the Internet, a culture of free enjoyment of products has been established since the very beginning, even more firmly for the Internet than for the television industry. Moreover, regarding audiovisual products, some important limits of the television industry, such as the dependence of time, are overcome by the Internet.

In the last few years, public broadcasters, as well as the private ones, have placed their contents in the Internet. This has been a gradual process and not exempt from doubts and difficulties for the television channels. Simultaneously, there has been a rapid increase in the use of the Internet. This reality brings about important issues related to the television business model and distribution of content. The Internet can be a threat to television channels or it can become an opportunity for them [DEN 03].

This explains why broadcasters have seen the strategic importance of the Internet, since television and the Internet develop a symbiotic relationship with significant financial implications. As Chan-Olmsted and Ha pointed out in 2003, the Internet was first used to enhance the core product of established businesses [CHA 03]. However, and following Doyle, the big question for broadcasters to answer is: "Is the migration to multi-platform enabling television companies to use their resources more effectively than before?" [DOY 10].

8.2. Broadcasters and online activities

As television consumption has changed, the first question to be answered is how to monetize online audiences [ZUC 08]. Against the disruption theory and "the displacement effect" [JAM 95], some authors see online consumption as a complement to television viewing. From a media economics perspective, it seems clear that there are reasons enough to believe in the multiple synergies between broadcasters and online video delivery [MED 15] [HA 06]. Engagement and branding seem to be the most valuable benefits that have to be monetized. Where there are opportunities for additional consumption and additional audience value, there are ways to monetize.

Therefore, there are new sources for revenues and new audience data are needed. If there are multiple ways in which audiences can access the same content, traditional audience measurement should change [BER 09] [TAN 12]. Laptops, tablet devices, smartphones and smart TVs are alternative screens which allow audience mobility [SCH 12] [DIE 14] [HAS 14] and generate "place shifting" [BUZ 12]. Besides, we can view programs at any time, and the traditional meters, which "took into account the program's hour of emission in order to measure the audience" [POR 15], are not useful any more.

Table 8.1 shows how a broadcaster monetized its services and products in the online strategy.

	Advertising	Data based			Direct payment
		Commercialization of content	Without registration	With registration or subscription	
Business strategies	– Ads in webs and digital platforms – Look for high audiences giving free access and promoting through social media – Programmatic sale – Ads in online videos	Offer born out content in other platforms	– Offer relevant content to keep users and obtain new ones – Offer online video related to content interesting for advertisers		– Payment from subscribers – Payment from pay per view
Targets	– Users of the online platforms – Video users	Video users		– Registered users – Subscribers	
Data available	– Visitors – Online activity	– Visitors – Online activity – Downloads – Valuable content		– Profiles – Online activity – Downloads – Valuable content	

Table 8.1. *Strategies for online video monetization*

The revenues from online video can come from nine different business strategies that have to deal with four targets. For making recommendations of content, media requires a continuous and automatic analysis of data. It requires the management of big data, and hence the task is really challenging. Alliances with technological companies, expert in consumer data management, geolocation and big data, seem to be necessary.

However, some authors show their doubts about the capability of new media to develop their own business independently of traditional television. According to [MUR 11], "television is central to new media". [HAR 02] questions whether "new media can find its own voice, its own vision, detached from the traditions of the cinematic and the televisual".

In the audiovisual landscape, most television companies have reacted to this threat by developing online strategies to experiment with new media. For example, some television shows have been launched first on the Internet through social media, and then on the screen; news programs have incorporated content generated by viewers, news coming from citizens; some shows have reached their finale with a last episode widely advertised through social media; others have been specifically produced for online media and have their own *YouTube* channel.

The question of where revenues stem from is not an easy one [PIC 09]. How to monetize these actions is the key question for media managers, since new media have different business models from the traditional ones. Therefore, these experiments may become an innovative source of income, different from traditional audience ratings and thirty-second spots.

Boyd and Ellison [BOY 08] define social media as "web-based services that allow users to build a public or semi-public profile within a system; articulate a user list with shared relationships; and view the list of relationships of those persons with other people within the system".

Social media are characterized by peer collaboration, transparency, audience participation, segmentation and globalization [TAP 06] [KOZ 06]. Some of these features are complementary to broadcast media, and thus, they can enrich their potential. However, new difficulties arise, such as piracy and copyright control, and media companies have to develop tools in order to avoid them or make the most of them [LI 08].

The basic economic characteristics that have been outlined in previous sections are at the root of whether or not the use of social media by broadcasters is profitable. The difficulty of charging for content lies in the nature of public goods as media products, and following Hayes and Graybeal, "by not charging for content, media providers have created a completely price elastic market for content" [HAY 11]. In such a market, substitution of products is highly predominant, and also the possibility of sharing content among users without cost. Aware of the potential of sharing content and the impossibility of stopping this practice, Hayes and Graybeal suggest a micropayment model that lies in the concept of sharing. In such a model, "sharing content with a friend would be exactly what the media provider would want consumers to do" [HAY 11]. In addition, that is exactly what social media encourage users to do.

As a consequence of the windowing strategies, the Internet makes it possible to spread audiovisual products, to repeat their consumption and to increase the audience's knowledge of them. Programs distributed in this way have already been broadcast on television, and hence have become a new type of rerun, but using different media (mobile phones, portable media players), often sooner after the original broadcast than was done before. The Internet avoids the constraints of the traditional grid, not only by making it more flexible, depending on the wishes of the user, but also by rendering the media more flexible in terms of time and place [HER 09]. From the broadcasters' point of view, complex windowing techniques will play an increasingly important role in the exploitation of television content assets [DOY 10].

Technology allows enthusiasts to develop activities that go beyond viewing again or following the programs in television magazines. If fan or enthusiast activity came before the Internet, the Web provides new means of participation. Sharing is one of the key aspects of social media [GUE 11]. Apart from visiting the official websites, enthusiasts have created new sites, and visit those of other fans. The community feeling that arises from these relationships between people, who had not met before, shows the possibilities of engagement for audiovisual products. In this sense, not only has the product spread to other media, but the way in which members of the audience understand this product has significantly changed as well.

Obviously, some audiovisual content is more likely to engage with the audience and, therefore, more suitable for developing social media strategies. Stefanone, Lackaff and Rosen [STE 10] talk about the popularity of so-called "reality television" and the concomitant adoption of social networking sites (SNSs). In fact, their study links the consumption of a specific television genre with specific behavioral outcomes in new media. According to Godlewski and Perse, reality television could be defined as follows: "unscripted programmes that record real people as they live out events in their lives, as these events occur (…) Moreover, for some of these programmes, audience participation is a central agent to the shared experience or lived reality of the program" [GOD 10].

Although audience participation gives the channels a certain amount of power, Siapera believes that the broadcasters' activity on the Internet through the official websites gives them even more power, as they can in this way control their fans' activity, learn their preferences, promote the creation of communities and, therefore, strengthen audience loyalty for the following *off line* broadcasts [SIA 04].

Moreover, other studies show the importance of using social media for news programs and how SNSs help to promote their on-air programs, escalate viewership and boost popularity among viewers. Since modern viewers go online for their communication experiences, SNS characteristics provide television networks with great brand management and relationship development opportunities. According to Lin and Peña, "through the use of socioemotional messages, television network brands can facilitate optimized self-presentation, favourable conversations about the brands, and valuable relationships with consumers at both individual and collective levels" [LIN 11].

The use of social media linked to programs is also very adequate for television events. Such events result in an increased willingness to tweet. For example, *Super Bowl* XLV produced more than 4,000 tweets per second during the final moments of the game [PAT 11]. According to data used by Patel and Kunur [PAT 11], some of the top programs in social media – such as Twitter, Facebook and other social networks – are not necessarily the top audience rating programs. Only in week 4–10 April 2011 was there a program that reached number one in both lists: *American Idol*.

8.3. The new online business

According to AMETIC [AME 11], thanks to the free access to social media, the main business model followed by social networks is *advertising*. The main added value of social networks lies in the number of users, and the time spent on them. These parameters are essential for advertisers to decide on which social network their campaigns will be launched [ECH 13] [ELL 10]. According to several studies, the primary social network is Facebook. It could surpass internet giants like Google or Yahoo in advertising revenue. Facebook's display advertising revenue will grow from $1.8 billion in

2010 to $3.5 billion expected in 2011. Meanwhile, Google (and dependent pages like *YouTube*) expected $2.6 billion in 2011, compared with a $2 billion revenue in 2010 [END 11]. Google has launched Google +, a social network of similar characteristics, to compete against Facebook, but it is currently not too successful [O′NE 10].

Another business model being used by social media is the *payment for the purchase* of virtual objects. In April 2010, Facebook launched *Facebook Credits*, a virtual currency for financial transactions made in the network. These Facebook Credits are purchased with real money (through credit card or PayPal) and can be used in more than 60 applications including the social network. It is, therefore, a replica of Freemium, used mainly in the gaming industry. Revenues from this model represent 10% of the total Facebook revenues.

The third model is *subscription or premium payment* for services. In the case of LinkedIn, premium accounts give users access to more information about searched profiles, knowledge about other premium users, etc. LinkedIn has several types of Premium accounts with subscriptions from $19.95/month to $74.95/month. According to company figures, in the first nine months of 2010, revenues from this business model accounted for 27% of the total revenues of the social network [PAR 11]. However, which of these is the most profitable business model is still not clear.

The fourth source of revenue comes from the *data* provided by online consumption. It is information with great value for internal decisions but also other activities such as direct marketing action for advertisers. Broadcasters need to know the value of data from users of their online platforms, registered users and subscribers. The use of the media's data sets for other activities such as targeted marketing can increase revenues, amplifying their business. Media have to learn how to manage valuable big data generated in their interactions with their online audience.

As we have seen, social media have proved to be useful means of establishing contact with the audience and fostering emotional engagement and brand loyalty. There is no doubt that they are both very valuable intangible assets for broadcasters.

In this line, some authors use the term "social capital" [BUR 05], which can be useful for our final considerations. Social capital is described as "the benefits individuals derive from their social relationship and interactions: resources such as emotional support, exposure to diverse ideas, and access to non-redundant information" [ELL 10].

However, social capital is very difficult to monetize, even from users. What social media provide for media companies, rather than revenues, is traffic to their sites; as for other companies, social media are a very efficient decision-making prescriber of their products. In 2010, the number of social network users worldwide would have reached 1 billion with a penetration of 19% of the total population, seven points more than in 2009. In 2010, Spain had 10.1 million social network users, which is a penetration of approximately 43%. Over 80% of Spanish Internet users are social network users, and 20% use at least three networks [ORA 11].

The potential to provide traffic to broadcasters' sites may attract advertising, which is the traditional way to monetize in the broadcasting industry. The greater the audience, the greater the advertising expenditure, which should also mean: the more traffic, the more advertising expenditure in social media. However, advertising expenditure is not at its best.

We wonder what is understood by profitability in this new context. At the moment, profitability may have more to do with nearness, identification, satisfaction, entertainment experience and branding, than with massive audiences and big revenues. As Picard [PIC 09] outlines, social media help to interact, but it is necessary to understand why interaction is needed and how benefits are going to be measured. The posts, the log-ins and the mentions in the different social media can be measured, but the difficult part is to monetize them. Benefits can be measured not only in monetary terms, but also in terms of popularity and brand.

Therefore, and following the economic foundations outlined previously, social media represent for broadcasters the most evolved step of windowing strategies. While in the previous steps, these strategies allow profits to be maximized through different prices and time slots, social media increase engagement with and knowledge of the product, based on the concept of sharing, which has its roots in the nature of the product as public good.

Those windowing techniques will require the convergence between the broadcasting industry and other industries. Complementarities refer to the degree to which a bundle of goods provides more value than the separate consumption of those goods [BRA 96]. Convergence between different new media technologies can be viewed from this notion of complements. Dowling, Lechner and Thielman [DOW 98] propose that, in a "complementary convergence" situation, additional synergistic products may emerge to form a larger market (a complement view). We could assert that the traditional economies of scope go beyond the broadcasting industry.

In that sense, many Internet-related opportunities will require the integration of skills and capabilities that reside outside a traditional media company. This, in turn, will lead to structural changes in the media sector which may involve market boundaries. That explains the increase in mergers and acquisitions among large media companies as a consequence of the emergence of the Internet [SUL 10].

Following this idea, Chan-Olmsted, Lee and Kin, regarding the convergence with mobile phones, assert: "The broadcasters, with a keen appreciation of the mobility and personalization value provided by the mobile platform and of its compatibility and complementarity with their current offerings, focus on enhancing mobile television content, finding ways to monetize the content beyond advertising revenues, and integrating the strategic value of mobile and fixed television to make their overall product portfolio more competitive" [CHA 11].

In the same vein, and looking for ways to monetize within a "free" culture, Hayes and Graybeal point out: "the mobile environment offers micropayment a clean slate. Smart phones have ushered in a culture of paying, in fact, micropaying, for apps and games where the Internet fostered a 'free' culture" [HAY 11].

Convergence of industries and complementarities of products are very much linked to the brand communication of the broadcasters themselves as well as their products. Following Lin and Peña, "the economic advantages usually generated from strong brand relationships imply that it is important for television network brands to determine not only SNS distribution strength but also the degree and type of involvement that consumers experience through SNS" [LIN 11].

It seems clear that from a media economics perspective, there are reasons enough to believe in the multiple synergies between broadcasters and social media. Engagement and branding seem to be the most valuable benefits that have to be monetized; they are the real "social capital". In the end, what is more important for broadcasters is to create popular content and try to facilitate their consumption in the highest possible number of platforms, and therefore tagging is an important task in order to increase the visibility through search engines.

8.4. Bibliography

[AME 11] AMETIC, *Informe sobre la Industria de contenidos digitales*, Ametic, available at: http://www.ametic.es/media/Ou80/File/Informe_ContenidosDigitales 2011.pdf, 2011.

[BER 09] BERMEJO F., "Audience manufacture in historical perspective: from broadcasting to Google". *New Media & Society*, vol. 11, nos 1–2, pp. 133–154, 2009.

[BOY 08] BOYD D., ELLISON N., "Social network sites: definition, history and scholarship, *Journal of Computer-Mediated Communication*, vol. 13, pp. 210–230, 2008.

[BRA 96] BRANDENBURGER A.M., NALEBUFF B.J., *Co-opetition*, Currency/Doubleday, New York, 1996.

[BUR 05] BURT R., *Brokerage and Closure: An Introduction to Social Capital*, Oxford University Press, Oxford, 2005.

[BUZ 12] BUZZARD K., *Tracking the Audience. The Ratings Industry from Analog to Digital*, Routledge, New York, 2012.

[CHA 03] CHAN-OLMSTED S., HA L., "Internet business models for broadcasters. How television stations perceive and integrate the Internet", *Journal of Broadcasting & Electronic Media*, vol. 47, no. 4, pp. 597–617, 2003.

[CHA 11] CHAN-OLMSTED S., LEE S., KIM H., "Competitive strategies in Korean mobile television markets: a comparative analysis of mobile operators and television broadcasters", *International Journal on Media Management*, vol. 6, no. 1, pp. 77–93, 2011.

[DEN 03] DENNIS E., "Prospects for a big idea – is there a future for convergence?", *International Journal of Media Management*, vol. 5, no. 1, pp. 7–11, 2003.

[DIE 14] DIEGO P. *et al.*, "Televisión conectada en España: contenidos, pantallas y hábitos de visionado", *Revista Mediterránea de Comunicación: Mediterranean Journal of Communication*, vol. 5, no. 1, pp. 179–199, 2014.

[DOW 98] DOWLING M., LECHNER C., THIELMAN B., "Convergence – Innovation and change of market structures between television and online services", *Electronic Markets*, vol. 8, no. 4, pp. 31–35, 1998.

[DOY 10] DOYLE G., "From television to multi-platform: less from more or more for less?", *Convergence*, vol. 16, no. 4, pp. 431–449, 2010.

[DUN 90] DUNNET P. *The World Television Industry. An Economic Analysis*, Routledge, London, 1990.

[ECH 13] ECHEGARAY L., PEÑAFIEL C., "La utilización de las redes sociales como nuevas herramientas aplicadas al análisis de audiencia", *Trípodos*, vol. 33, pp. 157–172, 2013.

[ELL 10] ELLISON N., STEINFIELD C., LAMPE C., "Connection strategies: social capital implications of Facebook-enabled communication practices", *New Media & Society*, vol. 13, no. 6, pp. 873–892, 2010.

[END 11] ENDERS ANALYSIS, "Facebook's threat to Google", available at: http://www.endersanalysis.com/category/key-companies/facebook#, May 2011.

[FRI 00] FRITH S., "The black box: the value of television and the future of television research", *Screen*, vol. 41, no. 1, pp. 33–50, 2000.

[GOD 10] GODLEWSKI L.R., PERSE E.M., "Audience activity and reality television: identification, online activity, and satisfaction", *Communication Quarterly*, vol. 58, no. 2, pp. 148–169, 2010.

[GUE 11] GUERRERO E., "El ecosistema multiplataforma de los grupos televisivos españoles: los formatos de entretenimiento", *Comunicación y Hombre*, vol. 7, pp. 85–103, 2011.

[HA 06] HA L., GANAHL R. (eds), *Webcasting Worlwide: Business Models of an Emerging Global Medium*, Lawrence Erlbaum, Mahwah, 2006.

[HAR 02] HARRIES D., *The New Media Book*, BFI, London, 2002.

[HAS 14] HASSOUN D., "Tracking attentions: toward an analysis of simultaneous media use", *Television & New Media*, vol. 15, no. 4, pp. 271-288, 2014.

[HAY 11] HAYES J., GRAYBEAL G. "Synerging traditional media and the social web for monetization: a modified media micropayment model", *Journal of Media Business Studies*, vol. 8, no. 2, pp. 19–44, 2011.

[HER 09] HERRERO M., "The economics of series of audiovisual entertainment products", in MEDINA M. (ed.), *Creating, Producing and Selling TV Shows. The Case of the Most Popular Dramedies in Spain*, Formal Press/Media XXI, Lisbon, 2009.

[HOS 97] HOSKINS C., MCFADYEN S., FINN A., *Global Television and Film. An Introduction to the Economics of the Business*, Oxford University Press, New York, 1997.

[JAM 95] JAMES M., WOTRING C., FORREST E., "An exploratory study of the perceived benefits of electronic bulletin board use and their impact on other communication activities", *Journal of Broadcasting & Electronic Media*, vol. 39, no. 1, pp. 30–50, 1995.

[KOZ 06] KOZINETS R.V. "Click to connect: netnography and tribal advertising", *Journal of Advertising Research*, vol. 46, no. 3, pp. 279–288, September 2006.

[LI 08] LI C., BERNOFF J., *Groundswell: Winning in a World Transformed by Social Technologies*, Harvard Business Press, Boston, 2008.

[LIN 11] LIN J.-S., PEÑA J., "Are You Following Me? A content analysis of TV networks' brand communication on Twitter", *Journal of Interactive Advertising*, vol. 12, no. 1, pp. 17–29, 2011.

[MED 15] MEDINA M., "Modelos de negocio: gratuidad y pago por contenidos", in *La audiencia en la era digital*, Editorial Fragua, Madrid, 2015.

[MUR 11] MURPHY S.C., *How Television Invented New Media*, Rutgers University Press, New Brunswick, 2011.

[NAP 10] NAPOLI P.M., *Audience Evolution: New Technologies and the Transformation of the Media Audiences*, Columbia University Press, New York, 2010.

[NIE 84] NIETO A., *La prensa gratuita*, Eunsa, Pamplona, 1984.

[NIE 00] NIETO A., *Time and the Information Market: the Case of Spain*, Eunsa, Pamplona, 2000.

[O'NE 10] O'NEIL N., "Facebook Could Generate $300 Million In Credits Revenue Next Year, All Facebook", available at: http://www.allfacebook.com/facebook-could-generate-300-million-in-credits-revenue-next-year-2010-09, 23 September 2010.

[ORA 11] ORANGE FOUNDATION, Informe eEspaña, Fundación Orange, Madrid, available at: http://www.informeeespana.es/docs/eE2011.pdf, 2011.

[OWE 92] OWEN B.M., WILDMAN S.S., *Video Economics*, Harvard University Press, Cambridge, 1992.

[PAR 11] PARR B., "LinkedIn's IPO: An Overview", *Mashable Business,* available at: http://mashable.com/2011/01/28/linkedins-ipo-an-overview/, 28 January 2011.

[PAT 11] PATEL K., SLUTSKY I., "Is social media killing TV?", *Advertising Age*, vol. 82, no. 16, p. 4, 2011.

[PIC 89] PICARD R., *Media Economics*, Sage, London, 1989.

[PIC 09] PICARD R., "Blogs, tweets, social media, and the news business", *Nieman Reports*, pp. 10–12, Fall 2009.

[POR 07] PORTILLA I., "La medición de audiencias en televisión", in ARTERO J.P. *et al.* (eds), *La televisión en España. Informe 2007*, Deusto, Barcelona, 2007.

[POR 15] PORTILLA I., "Television audience measurement: proposals of the industry in the era of digitalization", *Trípodos*, vol. 36, pp. 75–92, 2015.

[SCH 12] SCHNEIDERMAN S., "Tapping into Tabletomics: A Tablet Research Exploration", *Viacom.blog,* available at: http://blog.viacom.com, 17 April 2012.

[SIA 04] SIAPERA E., "From coach potatoes to cybernauts? The expanding notion of the audience on TV channels' websites", *New Media & Society*, vol. 6, pp. 155–172, 2004.

[STE 10] STEFANONE M.A., LACKAFF D., ROSEN D., "The Relationship between Traditional Mass Media and 'Social Media': Reality Television as a Model for Social Network Site Behavior", *Journal of Broadcasting and Electronic Media*, vol. 54, no. 3, pp. 508–525, 2010.

[SUL 10] SULLIVAN D., JIANG Y., "Media convergence and the impact of the Internet on the M&A activity of large media companies", *Journal of Media Business Studies*, vol. 7, no. 4, pp. 21–40, 2010.

[TAN 12] TANEJA H., MAMORIA U., "Measuring media use across platforms: evolving audience information systems", *The International Journal on Media Management*, vol. 14, pp. 121–140, 2012.

[TAP 06] TAPSCOTT D., WILLIAMS A., *Wikinomics: How Mass Collaboration Changes Everything*, Portfolio Trade, New York, 2006.

[WEB 14] WEBSTER J.G., *The Marketplace of Attention: How Audiences Take Shape in a Digital Age*, MIT Press, Cambridge, 2014.

[ZUC 08] ZUCKER J., "A time for change: NATPE keynote address of NBC Universal president/CEO Jeff Zucker", available at: http://www.broadcastingcable.com/news/programming/time-change-natpe-keynote-address-nbc-universal-presidentceo-jeff-zucker/31555, 2008.

9

Implementing Innovation Strategy

Most media companies have developed technological innovation and multimedia departments. In Doyle's view [DOY 10], the biggest television companies have migrated to multiplatform, characterized by the introduction of "360-degree commissioning", the development of websites and other digital offerings. However, in many cases, innovation is an experimental inquiry, rather than a strategic one. Apparently, the reasons have to do less with profitability and more with the need to defend the core business that is being seriously threatened by new competitors [GAL 09].

Nevertheless, private broadcasters regularly compete with the public broadcasters to develop attractive digital services. That is why the European Commission expects public media to lead the move toward digitization of the industry [EUR 07] [EUR 97]. To help public service media (PSM) plan for the digital transformation, in October 2012, the European Broadcasting Union (EBU) launched *VISION2020*. The goal was to identify how PSM can remain indispensable to audiences and stakeholders in the years ahead [EBU 13]. Some authors demand the public companies to develop strategies in order to fulfill the needs of different segments of their audience and even individual necessities [LOW 08, HIL 00].

The implementation of innovation within the company involves challenges regarding corporative culture and professional routines, among others. It requires coordination with legacy media within the company and a new media worker profile able to develop new products and services. The

Chapter written by Mercedes MEDINA.

final challenge is to succeed in launching new services at the right time, minimizing costs and offering added value to audiences.

The digitization of the media industry is related to two important changes: the penetration of the Internet and its use as both an advertising tool and a content distributor, and the use of different mobile devices to access media content [FER 08, PAG 00]. One of the challenges for media companies is to be able to deliver the content through the maximum number of devices, not only television, but also the Web, and mobile devices [WIL 08, JAC 07]. It forces traditional media gatekeepers to work harder at trying to determine what content or devices will succeed [LAW 05, p. 161].

The advent of convergence and the invention of new devices have caused both audience fragmentation and the transformation of media consumption [ALV 11]. Audiences are no longer passive but active, as they have the opportunity to choose between both linear and nonlinear programs, as well as to interact with the scheduling.

The convergence of Internet and TV has a particular form called IPTV. IPTV includes all video and ancillary services such as audio/text/data, delivered over the Internet Protocol, offered as a channel of linear and/or nonlinear programming of broadcast quality, and designed to be watched on a television set [OEC 07]. Catch-up, on the other hand, provides the opportunity to watch live TV channels, broadcast over the Internet and displayed on any computer screen. Mobile TV is the transmission of traditional and on-demand audiovisual content, to a capable mobile device, including live and time-shifted TV. It is therefore broader than broadcasting alone [EUR 07].

With Mintzberg [MIN 03], Küng [KUN 08a, p. 121] recognizes that the incorporation of the Internet is "one of the most challenging strategies for leaders and uncomfortable for those inside organizations, since it involves moving from the known to the unknown, and requires existing success formulas to be abandoned and new competences and attitudes to be developed". However, innovation always requires a change in the organization and the effort to achieve internal coordination.

Bearing in mind the difficulties of these processes of implementation and the need of media companies to accomplish them in the digital era, we wonder whether it is more efficient for a media organization to create a sister company to implement innovation and foster creativity, or to develop an internal department to do so. Though the deep question is what the media will become: will they be the gatekeepers and creators of quality content, or will they be just distributors of what others do? Hence, looking at this process, the key is to distinguish the permanent features from the changing ones.

9.1. Innovation culture

To understand the current innovation in media companies, it is necessary to take into account other aspects related to corporate culture and organization, human talent, and external alliances. Innovation, the act or process of creating something new, "enables organizations to improve the quality of their outputs, revitalize mature businesses, enter new markets, react to competitive encroachment, try out new technologies, leverage investment in technologies that are so expensive that no single product can recoup them, and develop alternative applications for existing product categories, to name a few outcomes. For organizations that must adapt to changing competition, markets, and technologies, product innovation is not a fad. It is a necessity" [DOU 96, p. 424].

According to Chan-Olmsted [CHA 05, p. 14], "strategy involves a range of a firm's decisions and activities that are enacted to fulfil the firm's strategic missions and goals through the effective use of skills and resources, considering the opportunities and threats in its market environment". In order to obtain that, Kachaner and Deimler [KAC 08] suggest that organizations need to stretch their strategic thinking by questioning and challenging the existing assumptions and business models. Briefly, implementing innovation is the process that has become one of the most important strategic decisions that media companies must take.

In order to implement any innovation successfully, it has to impact the structure of the organization, to be spread to the rest of the company, change the value chain, to be accepted by the target, to increase the quality of the products and services and to become profitable. More often, innovation also has legal implications.

In summary, the key factors to boost innovation successfully are related to the following items:

a) The number of employees contracted for the new divisions has to be large enough among the overall employees of the companies. This could be a sign that the firm has taken seriously the transformation of the company in a multiplatform strategy, diversifying its activity and modifying the core business.

b) Transformation requires alliances with external companies from the telecommunication and electronic sector.

c) Diversification might make sense whether revenues increase and this increment allows designing new revenue streams.

d) The strategy will be successful if new audiences are achieved and the current viewers use and consume the new services.

e) The transformation of a media company is authentic if it encourages the production and acquisition of new and quality content.

According to Christensen and Overdorf [CHR 00], there are three ways to develop a new business inside a company: (a) to spin out an independent company capable of developing the necessary values and process; (b) to create a new organizational structure within the corporate boundaries in which new processes can be developed and which is staffed by teams that are physically located together and individually charged with personal responsibility for the success of the project and (c) to acquire an organization whose processes and values closely watch the requirements of the new task. They argue that established firms tend to fail if all innovation is supposed to happen within the existing organization. New structures and new laboratories are necessary to facilitate innovation.

O'Hair *et al.* [O'HA 97] defined culture as the shared beliefs, values and practices of a group of people. A group's culture includes the language or languages used by group members as well as the norms and rules about how behavior can appropriately be displayed and how it should be understood. In addition to the corporate culture, the different professional styles need to be integrated. For example, Uljin [ULJ 00, p. 307] outlines the following common friction between engineers and marketing people: "the latter believe that scientists/engineers have no sense of time, costs, service, or competitive advantage. They hide in the lab and continue developing a product without strategic planning, holding standardization and technology sacrosanct, and expecting the client to adapt. The marketers are, in the eyes of the scientists/engineers, aggressive, demanding, and unrealistic. They want everything now, want to deliver a product before it is ready, are always in a hurry and impatient, or cannot decide what they want". Before the question of what has influenced more in a company, corporate or professional culture, Uljin [ULJ 00, p. 308] suggests that "a collective orientation toward corporate goals is difficult to realize because of the knowledge workers' need for their professional autonomy" and thinks that this professional autonomy will increase in the new companies.

How decisions such as innovation plans are carried out by those responsible for implementation has been examined from multiple perspectives. Bourgeois and Brodwin [BOU 84] identified five strategies that organizations can use when attempting to implement decisions. The *commander model* is employed when the organization makes a centralized decision, such as adopting a new innovation, and instructs or orders the rank and file to implement it. The *change model* approaches implementation efforts by using organizational structure and system properties, such as incentives and rewards as inducements for acquiescence. When implementation must be negotiated and sold at senior levels, a *collaborative model* is put into place. The *cultural model* is employed when the organization attempts to use the corporate or organizational culture as a means of success of implementation. Finally, a *crescive model* attempts to cultivate and nurture implementation among organizational members.

One of the difficulties of the innovation is to adapt the workers' routines to the new models of production, commercialization and distribution. As Oliver [OLI 14] pointed out, "a culture of adaptability and change in media organizations is the only basis on which emergent strategies can be effectively implemented". Most studies conclude that information sharing is a powerful tool for desensitizing employees to change. Lewis and Seibold [LEW 98] conclude that communication is at the heart of change adoption and implementation and organizations would do well to employ communication strategies that emphasize information sharing, feedback, employee participation and influence, and collaboration. However, we agree with Howkins [HOW 01, p. 213] when he says that "the raw material is human talent: the talent to have new and original ideas and to turn those ideas into economic capital and saleable products (...) The most valuable currency is not money but ideas and intellectual property, which are intangible and highly mobile". According to Küng [KUN 08a, p. 149], "the successful introduction of new programs, products and services depends on a person or team having a good idea and then being able to develop that idea further".

There are different reasons that explain why media companies should transform and offer online services: to strengthen their position on the market against the threat of the new media [MED 02]; as a source of revenue; as a consequence of technological developments that suggest a natural extension of the core business; as a way to reach audiences that could not be served cost-effectively before and to distribute content that was not made public before because the cost was too high; as a way to give better services to audience and to know the viewers better; and finally, in the case of public broadcasting services, as a form of maintaining its legitimacy in a new competitive environment. According to Oliver [OLI 14], ideas about alternative revenue models, piracy, audience fragmentation and industry consolidation in the media are strategic issues that need to be addressed, assumptions challenged and strategic responses to be made. In most of the cases, revenues coming from Internet services are not high enough; however, it is an efficient medium to reduce time and costs of operations, and "enabled networks of collaboration between distributors, customers and suppliers" [KUN 08b, p. 39].

Bakker and Sádaba [BAK 08] suggest that the following aspects should be included in the online services: responsiveness to the user, which refers to the possibility of contacting the editor, one-to-many communication, interaction between users (chat rooms or interactive message boards); the customization of content: user ability to adapt the content to his/her own needs; and immediacy, which refers to the frequency with which the site is updated.

Online content must be multimedia, to the extent to which text, graphics, sound, voice and (still and moving) images are translated and integrated into one digital form, all of which require internal coordination. "The intrinsic functionalities of the Internet – unlimited space and the possibility to combine text, graphics and video – meant stories could be handled in greater depth and richness" [KUN 08a, p. 177].

9.2. Media transformation

We would like to explore how traditional media companies have transformed into online multimedia companies in Spain, where the Internet is understood as a new medium with its own strategy and business model.

A good example of a small public company was the *Corporació Catalana de Mitjans Audiovisuals* – CCMA (Catalan Audiovisual Media Corporation) and, more specifically, the creation of *CCRTV Interactiva* (CCRTV Interactive), a sister company that produces and distributes content generated by the CCMA radio and TV channels. It seems to be a paradigm of media transformation, not only because it has developed Internet and multimedia content for CCMA, but also because the whole company mentality changed thanks to CCRTV Interactiva work. As a result, according to Prado and Fernández [PRA 06], it played a pioneering role in the new process of convergence in the region.

The Catalan public broadcasting service was created in 1983 as *Corporació Catalana de Ràdio i Televisió – CCRTV* (Catalan Radio and Television Corporation), and followed the Spanish model of public broadcasting with a high level of governmental intervention and a dual funding system based on subsidies and advertising. An important change took place in 2007. The CCRTV lost the words of "Radio and Television" of its corporate name and became CCMA. Behind this nominal change was the idea to enlarge the scope of its mission to new digital media.

The focus of the *CCRTV Interactiva* was established as the exploitation and commercialization of interactive services for TVC and Catalunya Radio, which made a direct and personal relationship with users possible. The objectives were to foster the development of Information Society in Catalonia; to create value with the available content within the Corporation, and develop more on-demand content promoting the participation of the audience.

In order to achieve these goals, the *CCRTV Interactiva* management established the following operative objectives [MIR 08]:

a) positioning CCMA as one of the leaders of delivering content and services through Internet and new devices;

b) making the content, brands and star system the most demanded and consumed in Catalan language on the new platforms;

c) becoming an on-demand company, beyond linear and traditional consumption of TV and radio programs and be able to provide multiple content to any outdoor device;

d) doing whatever is needed to know the users, to learn about them and their needs and likes, by developing trustful, rich and frequent relationships with them on a one-to-one basis;

e) creating added value thanks to new business models and new audiences, especially attracting young people.

As far as labor resources are concerned, *CCRTV Interactiva* started with 34 employees and in 2009 had a workforce of around one hundred, of which around 40 were external employees [MED 11b]. According to the projects and the amount of work, *CCRTV Interactiva* hired external companies. Since 2007, most of the services, such as technical, editorial and content management, were being subcontracted.

Regarding work organization, *CCRTV Interactiva*'s daily work structure was organized around teams of three people, which are in charge of a product. The company was structured around five divisions: technology, economic control, human resources, content and commercial. The content department required a larger number of workers than the rest. This department was the driver of the production, because it contained the know-how and the editorial mark of the company. The professional profiles are diverse according to the tasks they have to develop. The professional profiles were multimedia journalists, creative personnel and designers for the different platforms, computer programmers specialized on Web, media interfaces, Java and server-side technology, and online advertising salesmen.

One of the main challenges for the manager team was to deal with three different working cultures: technological, creative and commercial, but, at the same time, the key success factor of the company was to achieve a good blend of the three skills. They searched professional profiles that could integrate with the others. For example, online journalists who had on understanding of or were interested in technology, were up to date with new languages and features; young people and digital native, with hybrid profiles, such as designers interested in the editorial side; and finally, flexible people, who were able to work in teams.

However, even more difficult was to change the corporate culture of the CCMA legacy radio and television channels. Managers, content producers and journalists had to produce for the new platforms and not only for the television or radio channel. It was necessary to change from one-media conception to a multi-media content design. *CCRTV Interactiva* had to coordinate traditional media with new media. For this purpose, they created a *Multichannel Commission,* headed by the Director of Strategy and Brand of the CCMA. This Commission had to approve the guidelines of new products and their content and established the production.

At the beginning, the initiative of new ideas and projects usually came from either content creators or the online salesmen who tried to develop commercial agreements with any potential client. After some years, the areas of radio and television started to suggest new content to be developed online.

One of the best examples of a 360° content was the documentary about Picasso's Guernica (www.tv3.cat/30minuts/guernica) with videos, history, games and details of the work.

As far as economic indicators are concerned, although *CCRTV Interactiva* was not in the business for getting profits because it belongs to a public corporation, it has increased their revenues from Internet and mobile services, especially since they open the video on-demand service to the public free of charge in 2007 (Table 9.1).

	2003	2004	2005	2006	2007
Mobile	140	190	200	270	320
Internet	420	590	610	640	970
TOTAL	**560**	**780**	**810**	**910**	**1,290**

Table 9.1. *Mobile services and Internet revenue (000 €) (source: Miralles [MIR 08])*

The company has been actively marketing its online advertising space and developing new business models in order to minimize the financial burden that means to the consolidated group, and also to prove that the future of media is through these new platforms, also in terms of revenue streams. The revenue came from: teletexto advertising, advertising on Internet, broadband services, interactive services and web services. Expenses were covered by a subsidy from the Catalan government. In exchange for this funding, the *CCRTV Interactiva* had to achieve two indicators: a certain amount of a number of yearly pages viewed on all the CCMA sites and a minimum number of hours of audiovisual CCMA content download by users.

From a point of view of results, apart from the audience data, *CCRTV Interactiva* has been recognized by national and international organizations. They received more than 40 awards since 2002 [CCM 06].

We would like to outline other examples of innovation strategies of other media companies in Spain: RTVE and Atresmedia.

As for its online development, through the Department of Planning and Technological Innovation, RTVE opted for innovating around its presence in social networks, developing applications for smartphones and tablets (+24, rtve.es, Clan, +TVE), or the tool *botón rojo* (red button) to access all interactive content and services from smart TVs [MED 15]. Through their applications and social networks, users can share content, interact with programs or capture 30 s of live programs. Through their website, users can access all their current and archival programs, and see the popularity of downloaded videos. It has also developed a news service 24/7 via the Web and mobile app *24 horas*.

To promote innovation, in June 2009, RTVE launched the International Award "INVI" for Online Audiovisual Innovation to look for and encourage talent on the Web [MED 11a]. It also developed a Media Lab, as we saw before in Chapter 4, to experiment with new transmedia products and explore new narratives. For example, they developed a map of characters and a monograph of the battles occurred until the conquest of Granada related to "Isabel". Moreover, the series "Águila Roja" has created games and apps for mobile devices [HER 11, COS 12].

RTVE effort to follow some recommendations of the EBU and its documents "*Vision 2020. A EBU Project*" can be noted. Specifically, we can observe an effort for better understanding the audiences, deliver high-quality content, connect with the networked society and access audience feedback, continuous innovation and development and try to be relevant and trusted source of information and improve multiplatform strategy.

Finally, we can outline the activity of Antena 3, a private company. In order to stay ahead of new technology, Antena 3 put a great amount of energy into the development of multimedia. The Multimedia Division contracted 50 employees in 2007, which was around 2% of the overall employment, and launched different *online services*: *Tuclip.com*, a platform for viewers to submit videos and *tucanaltv*, composed of the finest videos from Antena 3, including news, humor and entertainment. Antena 3 was also the first television company in Spain (as well as the second in Europe following the BBC) to launch a channel delivered through *YouTube*. Since 2010, all these services were moved to the new site of Antena 3 and a *Community Manager* was incorporated to manage its content and websites. On the site, we can watch both catch-up television and entire TV seasons on demand, using their "salon [sitting room] mode". Antena 3 inserts advertising spots before each episode and six other spots where the episode is divided.

Since 2009, Antena 3 has developed *strategy 3.0*, with the intention of delivering content through the three platforms: television, Internet and smart phones. With this strategy, they tried to show their viewers not merely that they were a television channel, but also that they moved into a truly multimedia company with interactive services. Social media sites became new tools to enable audience interaction. Through Facebook, Windows Life, Tuenti and Twitter, audience members can address comments to the firm or share its content with other users. The company also created blogs related to their own channels, including one related to the audiovisual industry (*www.antena3.com/objetivotv/*), and several others related to general topics, such as fashion, current affairs and sports. To increase interactivity and audience participation, video-meetings was created to chat with actors, singers and writers, and to play online games related to the programs. Antena 3 also created its own room in the second life site *Habbo*.

When considering mobile phones as providers of audiovisual content, a number of differences become evident. For us, online entertainment is offered in free basics, while the telephone economy is based on payment by the user. It is necessary to make a distinction between mobile television and mobile video, although sometimes the two are interchangeable: mobile TV provides live television, while mobile video enables the download of videos on demand. In 2009, Antena 3 signed an alliance with Vodafone, the second largest telecommunication company to deliver their news, series and other programs in 3G mobile telephones for 1.5 € a week. It also signed an agreement intended to integrate Internet portals to mobile telephones in 3G devices with other telecom companies such as Movistar, Orange, Nokia, Nintendo, Sony and Microsoft.

The company created two apps for the mobile devices. In 2012, *Atresmediaconecta* appeared offering access to extra content related to television programs. In 2013, *Atresplayer* was launched giving access to all its television channels. This year the group changed its name from *Antena 3* to *Atresmedia*.

With the data of the users of the video platforms (registered or unregistered), *Atresmedia* can make recommendations of videos to users according to their interests, improving their experience and engagement. Users will continue consuming videos and could recommend the platform to other users, increasing the volume of the audience.

Besides the marketing strategy, all these data can also be used for promotional activities. *Atresmedia* offers a service of email marketing based on these data. The company offers to brands the possibility of sending messages to a well-defined segment of the users, keeping the property of the data. The segmentation could even consider content consumption. For example, if a group of the audience is interested in content related to travels, this target can be an interesting public for travel agencies. The other possibility was the integration of this information with others, as geolocation data generated by mobile devices. An app like *Atresplayer* can register location. If its users accept giving this information, *Atresmedia* could segment the audience by city, type of restaurant visited or other location data. The most difficult task for *Atresmedia* is the management of all these big data, especially when data come from its different online platforms.

Thus, the value of the data is becoming more relevant each day and it has not gone unnoticed. As a result, *Atresmedia* created the department *Atresdata* as an "audience finder" looking for higher affinity [POR 16].

9.3. Final thoughts

According to Levin [LEV 09], digital revolution transformation has been deeper that the industrial revolution. "The change is far more significant than the transition from radio to television or movies from larger screens to smaller screens. The latter developments did not alter the fundamental narrative structures, the underlying hierarchy of power, or the core foundations of basic business models. Media companies have transformed from a top-down business models to a bottom-up business model emerging out of the consumption behaviour of media users" [LEV 09, p. 258].

After studying the case of *CCRTV Interactiva*, we could conclude that, to implement innovation, it is necessary that the board of the organization is strongly committed with innovation. There is also a need to have a flexible corporative culture that boosts communication and coordination among companies within the corporation. To have a clear business orientation compatible with the public service remit seems capital in a media company.

It is obvious that digital media is the route to growth in the media industry. Making television and radio content "active" in new media is only the first step; the real challenge is to turn our organizations into 360°, that is to say, to develop a multiplatform creative factory. The challenge is to move all creative layers into new media. It is not only about technology, but also about intellectual work coming from writers, producers, directors, designers, cameramen, journalists and entertainers and to transform their work in new formats at all levels. One of the main strengths of a media company to become a multimedia interactive platform would be based on the archives of videos and other documents play a substantial role in the process of media transformation. The experience of many years dealing with audiences gives a deep knowledge of them and allows the creation of products and services focused on defined targets. To contract external work for specific projects, understanding technology and the ability to adapt your business faster than your competitors seem to be necessary too.

Finally, to implement innovation, the important thing is the growth of all key performance indicators: audience, revenue and knowledge of users. Legacy media were better building entry barriers. They accumulate experience, they are defended by legal protection and they have the capability of attracting a considerable amount of revenues. However, quite often protecting the past destroys innovation [PER 14]. Hence, legacy media companies need to put attention in the customer intelligence department, and in making the company a right place to attract talented people. Though new media and innovation supposes turmoil on the public mission of media companies and the nature of media, what are permanent values are searching for quality, because audiences´ perception of quality becomes crucial in the abundance economy and so the competitive market of media products. Therefore, one of the most important lessons for media companies is becoming learning organizations, focusing on audiences and trying to archive their tastes and needs [LOT 16]. In fact, "consumers are looking for accessing items rather than a legacy bundle: an article rather than a newspaper, a tune rather that a DVD, a film rather than linear TV" [DEP 14, p. 42].

Content quality helps to build a brand, and as Bellamy and Traudt [BEL 00, p. 157] outline, "brands are vital in expanding markets". Brand for new business is a means of establishing instant market credibility, and "is necessary in the leveraging of corporate assets to open new markets".

9.4. Bibliography

[ALV 11] ÁLVAREZ-MONZONCILLO J.M., *La televisión etiquetada: nuevas audiencias, nuevos medios*, Ariel, Barcelona, 2011.

[BAK 08] BAKKER P., SÁDABA, C., "The impact of internet on users", in KÜNG L. *et al.* (eds), *The Internet and the Mass Media*, Sage, Los Angeles, pp. 86–101, 2008.

[BEL 00] BELLAMY R., TRAUDT P., "Television branding as promotion", in EASTMAN S. (ed.), *Research in Media Promotion*, LEA, Mahwah, 2000.

[BOU 84] BOURGEOIS L.J., BRODWIN D.R., "Strategic implementation: five approaches to an elusive phenomenon", *Strategic Management Journal*, vol. 5, pp. 241–264, 1984.

[CCM 06] CCMA, Annual Report 2005, CCMA, Barcelona, available at: http://www.ccma.cat/doc/comptes_2005.pdf, 2006.

[CHA 05] CHAN-OLMSTED S., *Competitive Strategy for Media Firms*, Erlbaum, Mahwah, 2005.

[CHR 00] CHRISTENSEN C.M., OVERDORF M., "Meeting the challenge of disruptive innovation", *Harvard Business Review*, vol. 78, no. 2, pp. 67–76, March 2000.

[COS 12] COSTA C., PIÑEIRO T., "Nuevas narrativas audiovisuales: multiplataforma, crossmedia y transmedia. El caso de Águila Roja (RTVE)", *Icono14*, vol. 10, no. 2, pp. 6–28, 2012.

[DEP 14] DE PRATO G. *et al.* (eds), *Digital Media Worlds: The New Economy of Media*, Palgrave Macmillan, New York, 2014.

[DOU 96] DOUGHERTY D., "Organizing for innovation", in CLEGG S. *et al.* (eds), *Handbook of Organization Studies*, Sage, London, 1996.

[DOY 10] DOYLE G. "From Television to Multi-Platform. Less from More or More for Less", *Convergence. The International Journal of Research into New Media Technologies*, vol. 16, no. 4, pp. 1–9, 2010.

[EBU 13] EBU, Vision 2020. A EBU Project, Connecting to a Networked Society, EBU, 2013.

[EUR 07] EUROPEAN COMMISSION, An EU Strategy for Mobile TV, Memo 07/298, Brussels, 2007.

[EUR 97] EUROPEAN COMMISSION, Green Paper on the Convergence of the Telecommunications, Media and Information Technology Sectors, and the Implications for Regulations, COM 623, European Commission, Brussels, 1997.

[FER 08] FERNANDEZ A., DE MORAGAS M., Communication and Cultural Policies in Europe, Government of Catalonia, Barcelona, 2008.

[GAL 09] GÁLIK, M., "The lessons of IPTV services in Europe and in Hungary", *Media Interactivity: Economic and Managerial Issues International Workshop*, Neuchâtel, 2009.

[HER 11] HERNÁNDEZ M., GRANDÍO M., "Narrativa crossmedia en el discurso televisivo de Ciencia Ficción. Estudio de Battlestar", *Galactica* (2003–2010), vol. 28, pp. 1–20, 2011.

[HIL 00] HILLS J., MICHALIS M., "The internet: a challenge for public broadcasting?", *Gazette*, vol. 62, no. 6, pp. 477–493, 2000.

[HOW 01] HOWKINS J., *The Creative Economy. How People Make Money from Ideas*, Penguin, London, 2001.

[JAC 07] JAKUBOWICZ K., "Public service broadcasting in the 21st century. What chance for a new beginning?", in LOWE G.F., BARDOEL J. (eds.), *From Public Service Broadcasting to Public Service Media*, Göteborg, RIPE, Nordicom, 2007.

[KAC 08] KACHANER N., DEIMLER, "How leading companies are stretching their strategy", *Strategy and Leadership*, vol. 36, no. 4, pp. 40–43, 2008.

[KUN 08a] KÜNG L., *Strategic Management in the Media: Theory to Practice*, Sage, London, 2008.

[KUN 08b] KÜNG L., PICARD R., TOWSE R., *The Internet and the Mass Media*, Sage, Los Angeles, 2008.

[LAW 05] LAWSON-BORDERS G., *Media Organizations and Convergence: Case Studies of Media Convergence Pioneers*, Lawrence Erlbaum Associates, Mahwah, 2005.

[LEV 09] LEVIN J., "An industry perspective: calibrating the velocity of change", in HOLT J., PERREN A. (eds) *Media Industries. History, Theory, and Method*, Wiley-Blackwell, 2009.

[LEW 98] LEWIS L., SEIBOLD D., "Reconceptualizing organizational change implementation as a communication problem: a review of literature and research agenda", in ROLOFF M. (ed.), *Communication Yearbook*, Sage, Beverly Hills, 1998.

[LOT 16] LOTZ A., "The paradigmatic evolution of U.S. television and the emergence of internet-distributed television", *Icono14*, vol. 14, no. 2, pp. 112–142, 2016.

[LOW 08] LOWE G.F., "Customer differentiation and interaction for innovation: two CRM challenges for PSB", *Journal of Media Business Studies*, vol. 5, no. 2, pp. 1–23, 2008.

[MED 15] MEDINA M. et al., "Audience's behaviour and multiplatform strategies: the path towards connected TV in Spain", *Revista Austral de Comunicación*, vol. 4, no. 1, pp. 157–176, 2015.

[MED 11a] MEDINA M., OJER T., "La transformación de las televisiones públicas en servicios digitales en la BBC y RTVE", *Comunicar*, vol. 36, no. 18, pp. 87–94, 2011.

[MED 11b] MEDINA M. et al., "Implementation of innovation strategy: the case of CCRTV interactiva", in VUKANOVIC Z., FAUSTINO P. (eds), *Managing Media Economy, Media Content and Technology in the Age of Digital Convergence*, vol. XXI, Focal Press/Media, Lisbon, 2011.

[MED 02] MEDINA M., ARRESE A., "Competition between new and old media in economic and financial news markets", in PICARD R. (ed.), *Media Firms: Structures, Operations, Performance*, Erlbaum, Mahwah, 2002.

[MIN 03] MINTZBERG H. *et al.*, *The Strategy Process: Concepts, Contexts, Cases*, Prentice Hall, Harlow, 2003.

[MIR 08] MIRALLES S., "2008: the search for growth in the media industry", *EMMA Annual Conference*, Barcelona, IESE, February 9th, 2008.

[OEC 07] OECD, *IPTV: Market Developments and Regulatory Treatment*, OECD Digital Economy Papers, no. 137, OECD Publishing, Paris, 2007. DOI: http://dx.doi.org/10.1787/230651165186.

[O'HA 97] O'HAIR D., FRIEDRICH G., WIEMANN J. *et al.*, *Competent Communication*, St. Martin's Press, New York, 1997.

[OLI 14] OLIVER J.J., "Dynamic capabilities and superior firm performance in the UK media industry", *Journal of Media Business Studies*, vol. 11, no. 2, pp. 57–77, 2014.

[PAG 00] PAGANI M., *La TV nell'era digitale: le nuove frontiere tecnologiche e di marketing della comunicazione televisiva*, EGEA, Milan, 2000.

[PER 14] PÉREZ-LATRE F.J., SÁNCHEZ-TABERNERO A., *Innovation in the Media: The Road to Change*, Media XXI, Lisbon, 2014.

[POR 16] PORTILLA I., MEDINA M., "Monetization strategies and audience data for online video. The case of Atresmedia", *Quaderns del CAC*, vol. 42, no. 19, pp. 27–36, 2016.

[PRA 06] PRADO E., Fernández D., "The role of public service broadcasters in the era of convergence. A case study of the television de catalunya", *Communication and Strategies*, vol. 62, no. 2, pp. 49–69, 2006.

[PRE 01] PRESTON P., *Reshaping Communications Technology, Information and Social Change*, Sage Publishers, London, 2001.

[ULJ 00] ULJIN J. *et al.*, "Innovation, Corporate Strategy, and Cultural Context: What is the Mission for International Business Communication?", *Journal of Business Communication*, vol. 37, pp. 293–315, 2000.

[WIL 08] WILDMAN S., "Interactive channels and the challenge of content budgeting", *The International Journal on Media Management*, vol. 10, no. 3, pp. 91–101, 2008.

Index

K, L, M

N, O, P

Q, R, S

T, U, V

W, Y

Other titles from

iSTE

in

Innovation, Entrepreneurship and Management

2016

BARBAROUX Pierre, ATTOUR Amel, SCHENK Eric
Knowledge Management and Innovation (Smart Innovation Set – Volume 6)

BEN BOUHENI Faten, AMMI Chantal, LEVY Aldo
Banking Governance, Performance And Risk-Taking: Conventional Banks Vs Islamic Banks

BOUTILLIER Sophie, CARRÉ Denis, LEVRATTO Nadine
Entrepreneurial Ecosystems (Smart Innovation Set – Volume 2)

BOUTILLIER Sophie, UZUNIDIS Dimitri
The Entrepreneur (Smart Innovation Set – Volume 8)

BOUVARD Patricia, SUZANNE Hervé
Collective Intelligence Development in Business

GALLAUD Delphine, LAPERCHE Blandine
Circular Economy, Industrial Ecology and Short Supply Chains (Smart Innovation Set – Volume 4)

GUERRIER Claudine
Security and Privacy in the Digital Era (Innovation and Technology Set – Volume 1)

MEGHOUAR Hicham
Corporate Takeover Targets

MONINO Jean-Louis, SEDKAOUI Soraya
Big Data, Open Data and Data Development
(Smart Innovation Set – Volume 3)

MOREL Laure, LE ROUX Serge
Fab Labs: Innovative User
(Smart Innovation Set – Volume 5)

PICARD Fabienne, TANGUY Corinne
Innovations and Techno-ecological Transition (Smart Innovation Set – Volume 7)

SAVIGNAC Emmanuelle
The Gamification of Work: The Use of Games in the Workplace

2015

CASADELLA Vanessa, LIU Zeting, DIMITRI Uzunidis
Innovation Capabilities and Economic Development in Open Economies
(Smart Innovation Set – Volume 1)

CORSI Patrick, MORIN Dominique
Sequencing Apple's DNA

CORSI Patrick, NEAU Erwan
Innovation Capability Maturity Model

FAIVRE-TAVIGNOT Bénédicte
Social Business and Base of the Pyramid

GODÉ Cécile
Team Coordination in Extreme Environments

MAILLARD Pierre
Competitive Quality and Innovation

MASSOTTE Pierre, CORSI Patrick
Operationalizing Sustainability

MASSOTTE Pierre, CORSI Patrick
Sustainability Calling

2014

DUBÉ Jean, LEGROS Diègo
Spatial Econometrics Using Microdata

LESCA Humbert, LESCA Nicolas
Strategic Decisions and Weak Signals

2013

HABART-CORLOSQUET Marine, JANSSEN Jacques, MANCA Raimondo
VaR Methodology for Non-Gaussian Finance

2012

DAL PONT Jean-Pierre
Process Engineering and Industrial Management

MAILLARD Pierre
Competitive Quality Strategies

POMEROL Jean-Charles
Decision-Making and Action

SZYLAR Christian
UCITS Handbook

2011

LESCA Nicolas
Environmental Scanning and Sustainable Development

LESCA Nicolas, LESCA Humbert
Weak Signals for Strategic Intelligence: Anticipation Tool for Managers

MERCIER-LAURENT Eunika
Innovation Ecosystems

2010

SZYLAR Christian
Risk Management under UCITS III/IV

2009

COHEN Corine
Business Intelligence

ZANINETTI Jean-Marc
Sustainable Development in the USA

2008

CORSI Patrick, DULIEU Mike
The Marketing of Technology Intensive Products and Services

DZEVER Sam, JAUSSAUD Jacques, ANDREOSSO Bernadette
Evolving Corporate Structures and Cultures in Asia / Impact of Globalization

2007

AMMI Chantal
Global Consumer Behavior

2006

BOUGHZALA Imed, ERMINE Jean-Louis
Trends in Enterprise Knowledge Management

CORSI Patrick *et al.*
Innovation Engineering: the Power of Intangible Networks

Printed and bound by CPI Group (UK) Ltd, Croydon, CR0 4YY